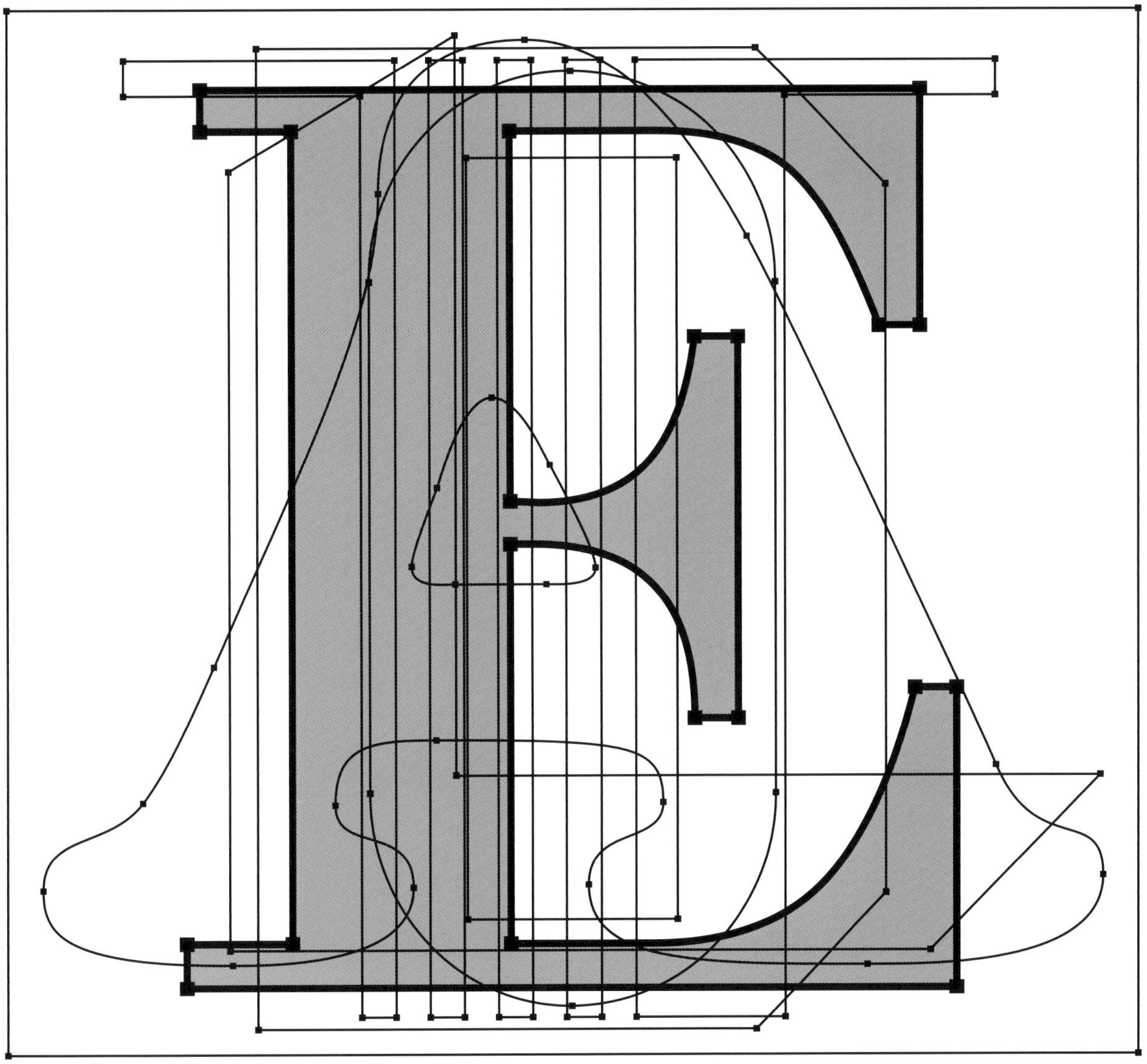

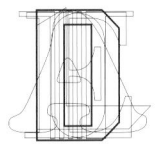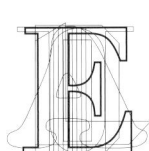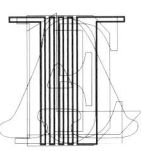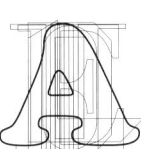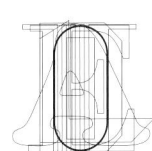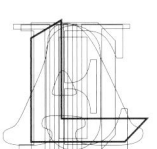

OFFICE·
EDUCATION

D-SERIES OFFICE / EDUCATION

OFFICE / EDUCATION 업무 / 교육

- 004 **HANA BANK MYEONGDONG BRANCH**   하나은행 명동지점
- 012 **KTB INVESTMENT & SECURITIES HEAD OFFICE**   KTB 투자증권 본사 사무공간
- 022 **COMING SPRING OF FLOWER**   꽃피는 봄이 오면
- 028 **ABILITY SYSTEMS**   어빌리티 시스템즈
- 034 **CROSS-STITCH**   크로스-스티치
- 040 **JAGOK-DONG OFFICE**   자곡동 사무실
- 044 **DMC CJ ENTERTAINMENT & MEDIA CENTER**   DMC CJ 엔터테인먼트 & 미디어 센터
- 056 **B2Y OFFICE REMODELING**   B2y 오피스 리모델링
- 072 **YOUNGONE HEAD OFFICE REMODELING**   영원무역 본사
- 080 **GREY GROUP**   그레이월드 와이드 코리아
- 086 **LS NETWORKS OFFICE**   LS 네트웍스
- 092 **IBK PRIVATE BANKING CENTER**   기업은행 강남 PB 센터
- 102 **ING SECURITIES BROKERAGE CO., LTD.**   ING PB 증권중개(주)
- 108 **HUB ENTERTAINMENT**   허브 엔터테인먼트
- 114 **DAIN LANDSCAPE OFFICE**   다인 조경 오피스
- 118 **ALPHA VILLE 44 PRODUCTION**   알파빌 44 프로덕션
- 126 **KTF GALLERY-THE ORANGE**   KTF 갤러리-오렌지
- 136 **AN ANNEX TO THE ASIA PUBLICATION CULTURE AND INFORMATION CENTER**   아시아 출판문화정보센터 별관
- 142 **LITTLE BEAR**   리틀 베어
- 150 **ORBI MATHEMATICS**   오르비 수학전문학원
- 154 **BAMBINI & KAGE**   BAMBINI 교육센터 & KAGE 영재교육학술원

HANA BANK MYEONGDONG BRANCH

(주)제이이즈워킹 | 장순각(한양대학교) jay is working. l Jang Soon Gak

하나은행 명동지점은 각종 간판과 다양한 색상, 질감의 파사드가 즐비한 명동 거리에 위치한다. 이러한 콘텍스트를 고려하여 간결하면서도 세련된 새로운 파사드를 계획하고자 했다. 마감재로 한국 전통 백자를 선택했다. 작가가 직접 200x200 유닛을 몰드로 뜨고, 가마에 구워 제작하는 방식으로 만들었다. 유닛 하나하나가 예술 작품이 되고, 유닛의 조합으로 만들어진 파사드는 명동 거리의 조형 예술 작품이 되었다. 유닛 사이사이에 프로그래밍된 언어에 따라 반응하는 LED 유닛을 배치했다. LED 유닛은 센서에 의해 건물 내부로 이동하는 사람들의 움직임을 읽어 반응하기도 한다. 이러한 건물과 사람 사이의 소통은 건물의 인지도와 활용도를 높인다. 파사드 하부 사선 방향으로 계획된 계단은 내부 계단으로 이어져 강한 축을 형성한다. 재료의 대비는 라운지와 서비스 공간으로의 동선을 나눈다. 외부에서부터 이어진 사선의 직선적 요소는 바닥에서 벽면으로 이어지며 곡선으로 전환되어 공간감을 극대화 시킨다. 모든 방문객은 Interactive Kit를 부여 받아 각 공간마다 숨겨진 센서로 미디어 Interaction을 경험하게 된다. 물리적인 3차원 공간이 아닌 미디어와 테크놀로지 그리고 감성이 함께 공존하는 4D 공간 체험이 된다.

Hana Bank Myeong-dong Branch is located on one of the busiest street that boasts colorful signboards and various textured facades. A simple and sophisticated facade was designed in consideration of the urban context and the traditional Korean white base was selected for its finish. The project was constructed of the 200x200 size units that were made by craftsmen who molded and baked the units in a kiln one by one. As each unit is an art, the facade made of the units becomes a huge art installation on the street in Myeongdong. Between the units, LED units that responds to the programmed language were placed. The LED units detect and respond to people's movement within the building through a sensor. Such communi-cations between architecture and users increase recognition and make better use of the space. The staircase placed on the bottom of the facade in the diagonal direction is connected to the interior stairways, forming an outstanding axis. The contrast of the materials divides a circulation into lounge and the service spaces. The diagonal lines drawn from the exterior are transformed into curves that optimize the spatial sense, connecting the floor with the walls. All visitors will experience media interaction through the given interactive kit and a sensor hidden in the space. Then they will feel as if they are in a 4D space where media, technology and emotions are integrated, not in a physical three - dimensional space.

위치 서울특별시 중구 명동 1가 65-2
면적 582.81m²
주요마감 외관 – 백자 타일, LED 타일, 대리석, 목재(이로코) / 바닥 – 세라믹, 목재(이로코) / 벽 – 자작나무 합판 위 코팅, 목재(이로코), 래커 / 천장- V.P. 도장, 자작나무 합판 위 코팅
설계기간 2010. 8 ~ 11
공사기간 2010. 9 ~ 12
디자인팀 서영지, 황용호, 이현주
공사팀 채원우, 최승철
사인디자인 송종현, 김경환
사진 남궁선

Location Hana Bank, 65-2, Myeong-dong 1-ga, Jung-gu, Seoul
Area 582.81m²
Finishing Facade - White ceramic tile, LED tile, Marble, Iroko / Floor - Ceramic, Iroko / Wall - Coating on birch plywood, Iroko, Lacquer / Ceiling - V.P, Coating on birch plywood
Design period 2010. 8 ~ 11
Construction period 2010. 9 ~ 12
Photographer Namgoong Sun

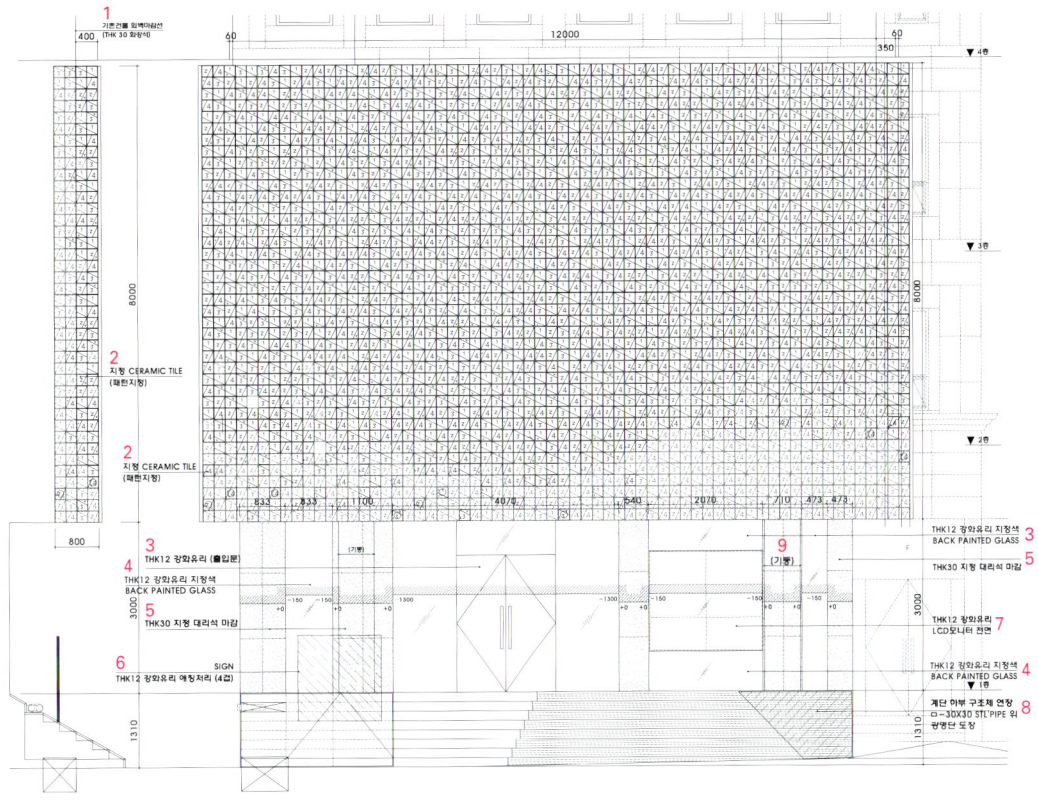

정면도 / front elevation

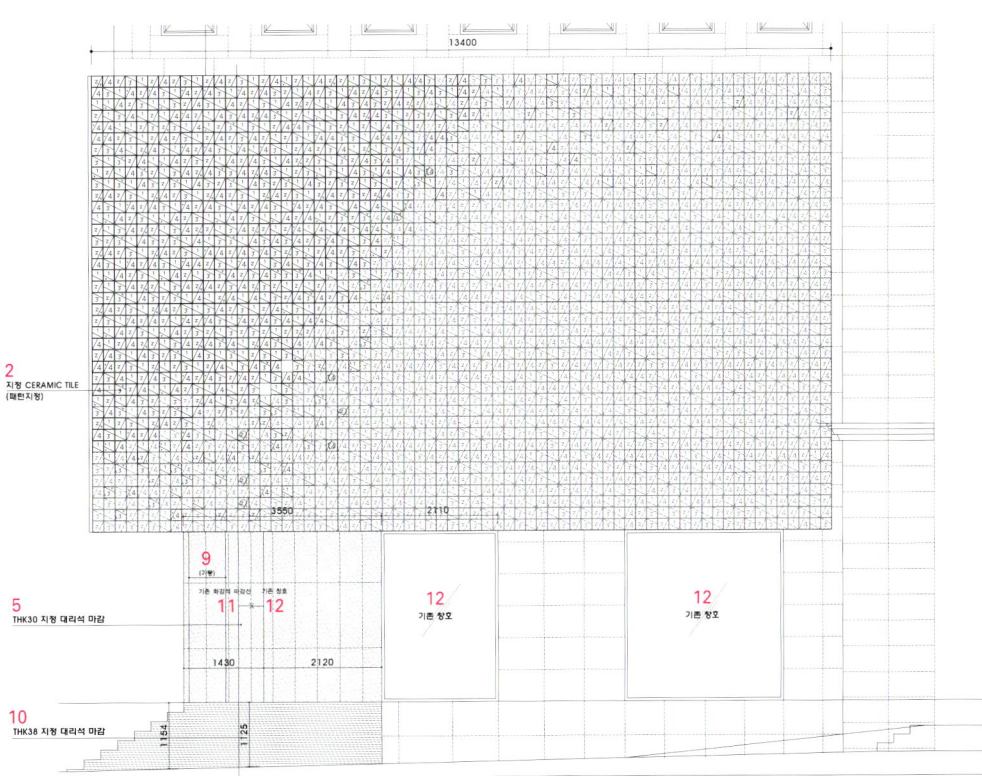

우측면도 / right elevation

1 EXISTING BUILDING OUTER WALL LINE(T30 GRANITE) 2 CERAMIC TILE(APP. PATTERN) 3 T12 TEMPERED GLASS(ENTRANCE DOOR) 4 T12 COLOR TEMPERED GLASS / BACK PAINTED GLASS 5 T30 MARBLE FIN. 6 SIGN / T12 TEMPERED GLASS ETCHING(4PLY) 7 T12 TEMPERED GLASS / LCD MONITOR 8 STAIR STRUCTURE EXTENSION / RED LEAD PAINTING ON □-30X30 STEEL PIPE 9 COLUMN 10 T38 MARBLE 11 EXISTING MARBLE FINISHING LINE 12 EXISTING WINDOW

1 기존 건물 외벽 마감선(T30 화강석) 2 지정 세라믹 타일(패턴 지정) 3 T12 강화 유리(출입문) 4 T12 강화 유리 지정색 / 백페인트 글라스 5 T30 대리석 마감 6 사인 / T12 강화 유리 에칭 처리(4겹) 7 T12 강화 유리 / LCD 모니터 전면 8 계단 하부 구조체 연장 / □-30X30 철제 파이프 위 광명단 도장 9 기둥 10 T38 대리석 마감 11 기존 화강석 마감선 12 기존 창호

1 GREEN COLOR W/PATTERN 2 WHITE COLOR CERAMIC 3 VOID 4 HOLE FOR HANGING 5 CERAMIC TYPE A 6 CERAMIC TYPE B

1 녹색 W/패턴 2 백색 세라믹 3 빈 공간 4 걸이 구멍 5 세라믹 타입 A 6 세라믹 타입 B

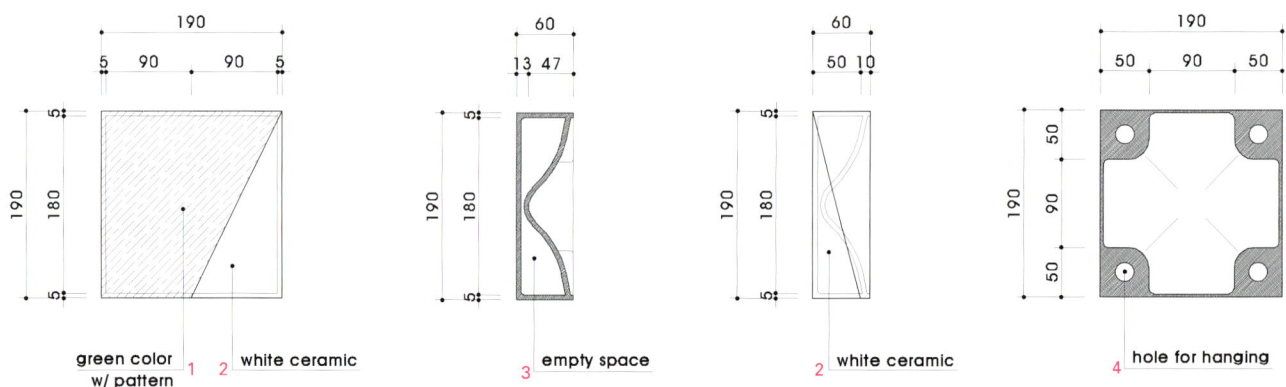

세라믹 유닛 타입 A / ceramic unit type A

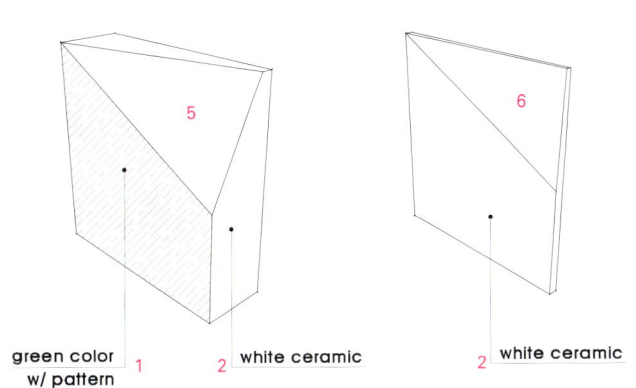

green color w/ pattern 1   2 white ceramic   2 white ceramic

세라믹 유닛 투시도 / ceramic unit perspective view

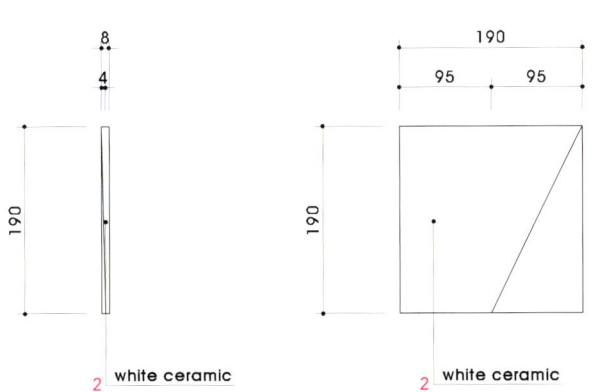

2 white ceramic   2 white ceramic

세라믹 유닛 타입 B / ceramic unit type B

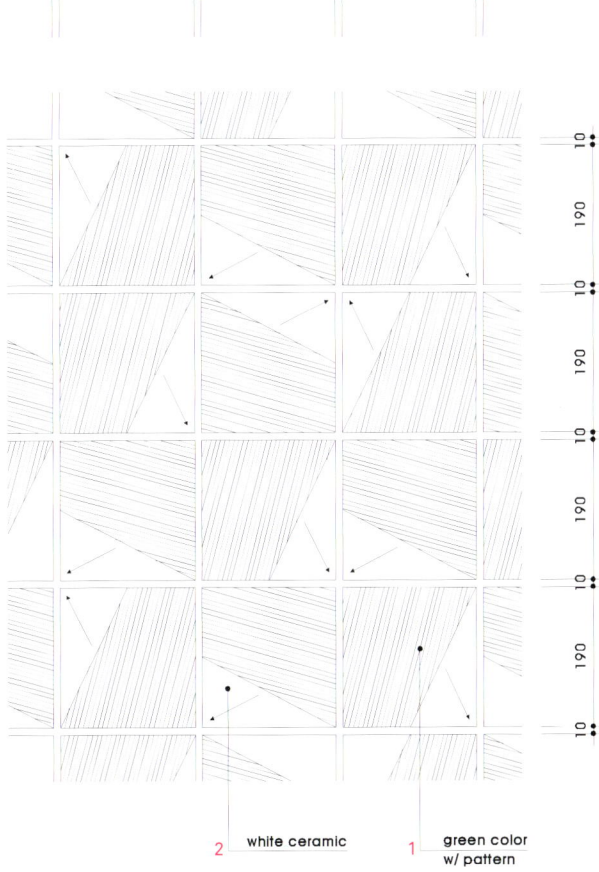

2 white ceramic   1 green color w/ pattern

유닛 조합 / unit combination

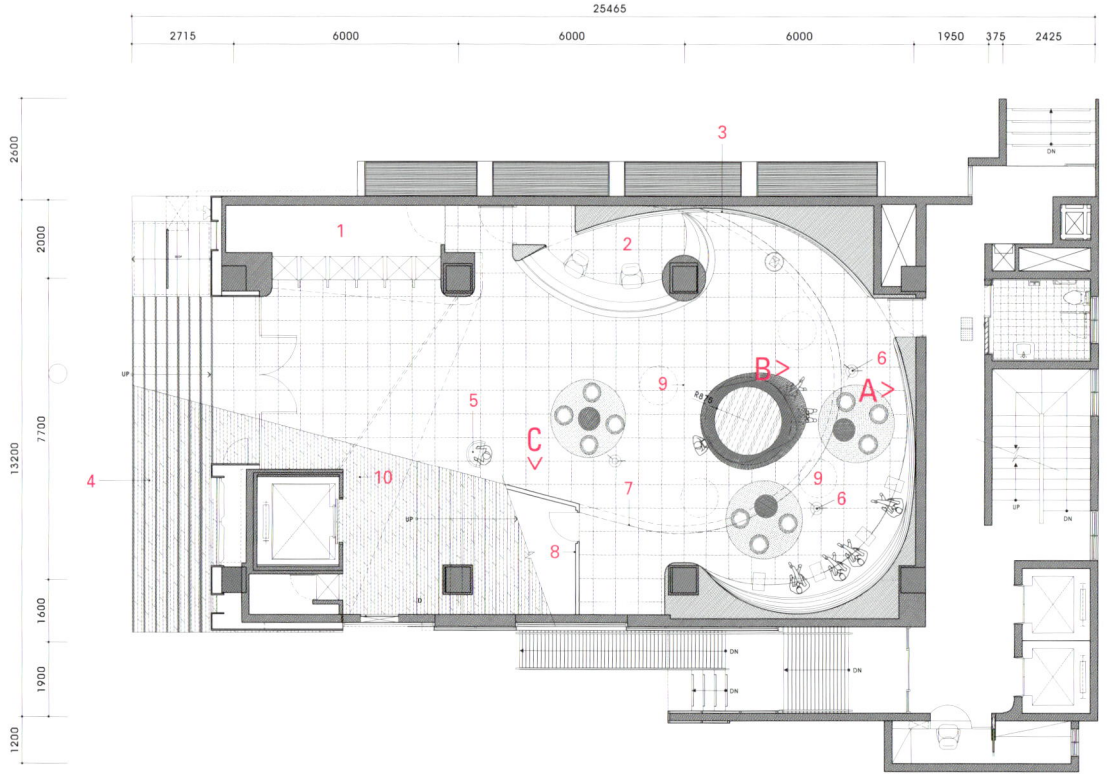

1 ATM **2** BOOTH FOR EXCHANGE **3** DONATION TREE **4** T38 WOOD FIN. **5** RECEPTION DESK **6** STAND TYPE LIGHTING **7** WOOD WAVE WALL **8** PHOTO MEDIA WALL **9** SPOT SOUND
10 HALL FLOOR / WOOD FIN. **11** CH 2,500 / WHITE PAINTING **12** OPEN TO 2ND FLOOR **13** CH 3,300 / WHITE PAINTING **14** CH 3,300 / BARRISOL ON CEILING BACKSIDE / LIGHTING
15 CH 4,300 / PAINTING ON EXPOSED CEILING

평면도 / floor plan

1 ATM실 2 환전 부스 3 도네이션 트리 4 T38 지정 목재 마감 5 안내 데스크 6 스탠드 조명 7 우드 웨이브 월 8 포토 미디어 월 9 스포트 사운드 10 홀 바닥 / 목재 마감 11 천장고 2,500 / 백색 도장 12 2층까지 열림 13 천장고 3,300 / 백색 도장 14 천장고 3,300 / 바리솔 천장 후면 / 전면 조명 15 천장고 4,300 / 노출 천장 위 도장 처리

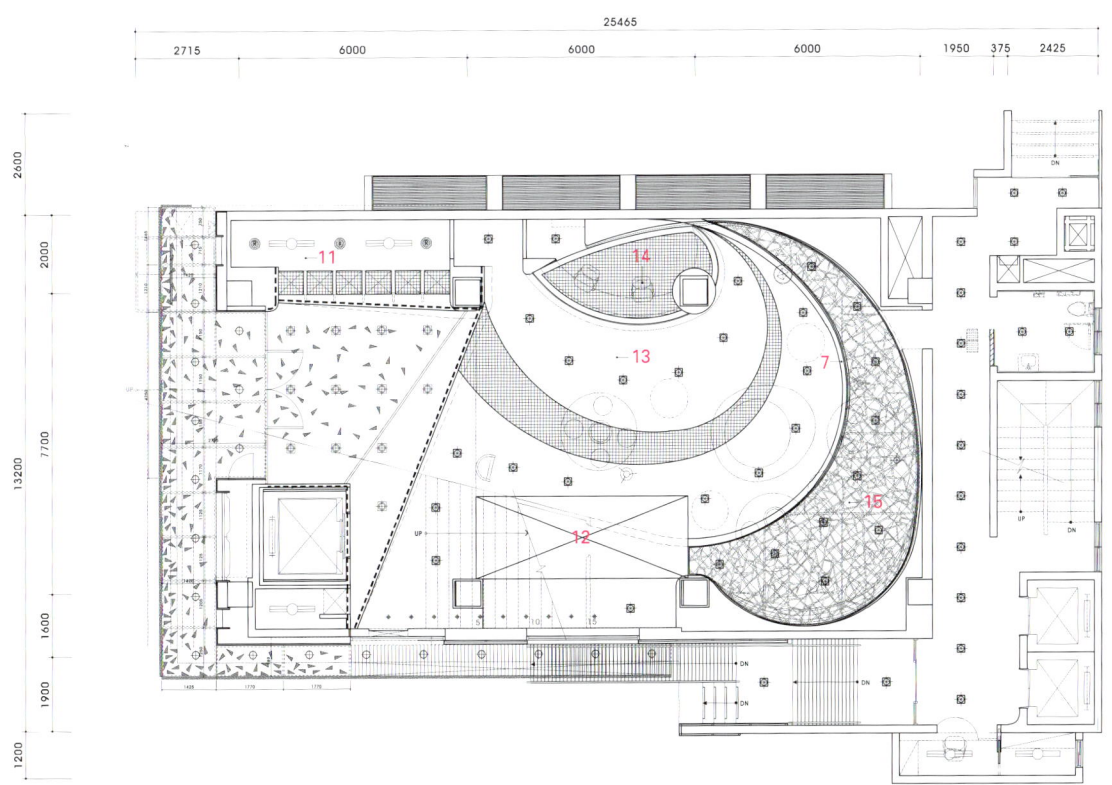

천장도 / ceiling plan

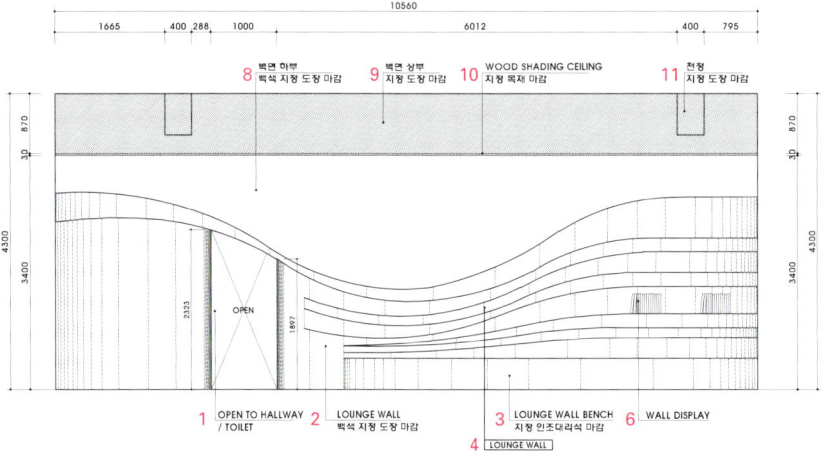

라운지 입면 A / lounge elevation A

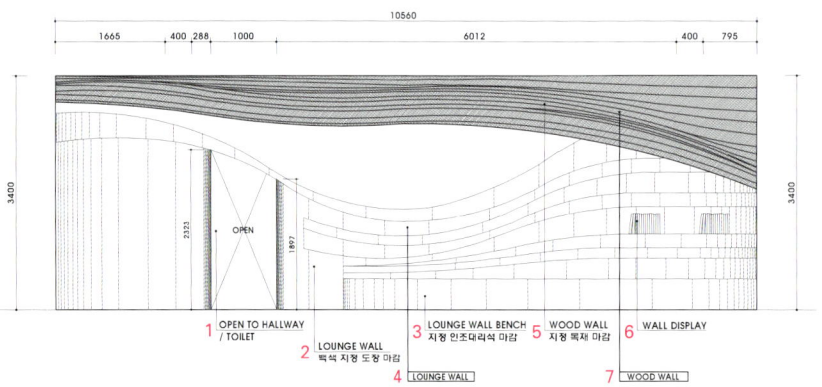

라운지 입면 B / lounge elevation B

1 OPEN TO HALLWAY / TOILET 2 LOUNGE WALL / WHITE PAINTING 3 LOUNGE WALL BENCH / 4 LOUNGE WALL 5 WOOD WALL / WOOD FIN. 6 WALL DISPLAY 7 WOOD WALL 8 WALL LOWER PART / WHITE PAINTING 9 WALL UPPER PART / PAINTING 10 WOOD SHADING CEILING / WOOD 11 CEILING / PAINTING 12 LOUNGE WALL / PAINTING 13 GESTURE WALL 14 IPONE STORAGE BOX 15 UPPER PART UPLIGHT 16 WALL / WHITE PAINTING 17 STAIR SIDE / WHITE PAINTING 18 STAIR SIDE / VENEER FIN. 19 STAIR SIDE / T12 TEMPERED GLASS 20 T12 APP. COLOR / BACK PAINTED GLASS 21 COLUMN / GRAPHIC PERFORATING ON WHITE PAINTING

1 홀 방향으로 열림 / 화장실 2 라운지 월 / 백색 지정 도장 마감 3 라운지 월 벤치 / 지정 인조 대리석 마감 4 라운지 월 5 우드 월 / 목재 마감 6 월 디스플레이 7 우드 월 8 벽면 하부 / 백색 도장 마감 9 벽면 상부 / 도장 마감 10 우드 쉐이드 천장 / 지정 목재 마감 11 천장 / 도장 마감 12 라운지 월 / 지정 도장 마감 13 제스처 월 14 IPONE 보관함 15 상부 업라이트 16 벽면 / 백색 도장 마감 17 계단 측면 / 백색 도장 마감 18 계단 옆면 / 무늬목 마감 19 계단 옆면 / T12 강화 유리 마감 20 T12 지정색 / 백페인트 글라스 21 기둥 / 백색 도장 마감 위 지정 그래픽 타공

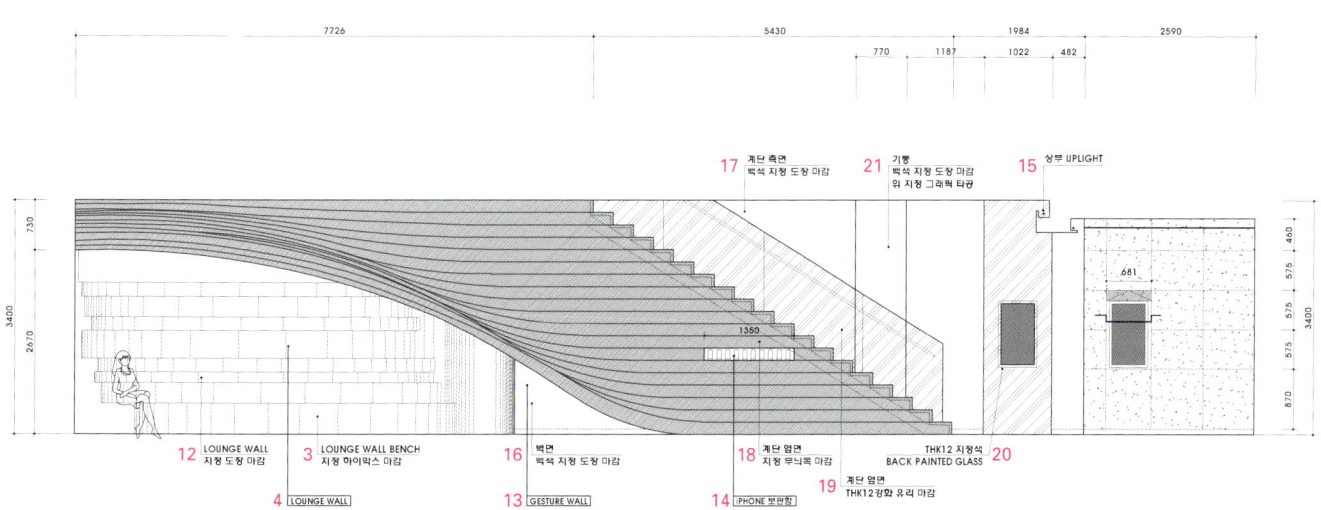

라운지 입면 C / lounge elevation C

KTB INVESTMENT & SECURITIES HEAD OFFICE

로디자인 | 김동진(홍익대학교), 김유정   L'EAU Design CO.,Ltd | Kim Dong Jin(Hongik University), Kim Yoo Jung

KTB 투자증권 본사 사무공간은 KTB 사무 혁신 프로젝트 중 하나이다. 다양한 색상의 종이들이 새로운 창작물을 만들어 내는 '색종이 접기'를 모티브로 공간 구성을 이끌어 냈다. 오피스의 기본적인 공간 틀은 다목적 셀이다. 정해진 용도의 '실'이 아니라 여러 가지 행위가 일어날 수 있는 '셀'들을 모듈화 하여 배치함으로써 다양한 성격의 공간이 복합적으로 구성된다. 각각의 셀들은 '커뮤니케이션 월'을 통하여 개방되거나 독립적인 공간이 된다. 이 벽은 공간을 구획하는 것뿐만 아니라 가구처럼 수납 공간으로 쓰이고, 벽에 누워 책을 볼 수 있는 등의 프로그램이 반영됐다. 또한, 혁신적인 의미를 지닌 작품을 선정하여 공간에 적용하는 '아트 피그먼트'(예술성을 흩뿌린다는 의미)를 통해 복도, 사무실 등의 한 공간에 배치시켜 하나의 연속적인 띠를 이루도록 했다. '색종이 접기' 콘셉트는 KTB 본사 사무공간뿐 아니라 계속적인 매뉴얼 작업으로 KTB 영업지점 공간에 진행된다.

As part of KTB's office innovation programme, the Yeoui-do KTB investment & securities head office has been redesigned. Based on the artistic concept of 'coloured - paper folding', the project is reconfigured as a new space in various combinations that respects the different colours of individual users. The basic frame work of the office space is multi - purpose cell. Instead of creating rooms with fixed uses, the project arranges modularised cells that can accommodate many different kinds of activities. As a result diverse spaces are generated as spaces with different characters are combining and compounding with each other. These new spaces serve as a place of free communication between different departments, encouraging diverse ideas to flow. Each cell becomes as open or closed space by means of the 'communication wall', which accommodates various programmes such as storage space as furniture or a space for reading books while lying down inside it in addition to the usual division of space. The spaces are connected via 'art pigments', which implies the spraying of artistic quality. These are then arranged as part of corridor or office to form a continuous spatial link to provide connections between the spaces with heterogeneous qualities. The Concept of 'coloured - paper folding' is developed not only for the head office but for the overall concept that will guide the design of all KTB offices in the future.

위치 서울특별시 영등포구 여의도동 신한금융투자타워
용도 업무
면적 1,175m²
완공 2010. 11
설계팀 강지혜, 이상학, 백소원, 김태연
시공팀 정구민
건축주 KTB 투자증권
사진 염승훈

Location Shinhan Investment Corporation Tower, Yeouido-dong, Yeongdeungpo-gu, Seoul
Use Office
Area 1,175m²
Completion 2010. 11
Photographer Yum Seoung Hoon

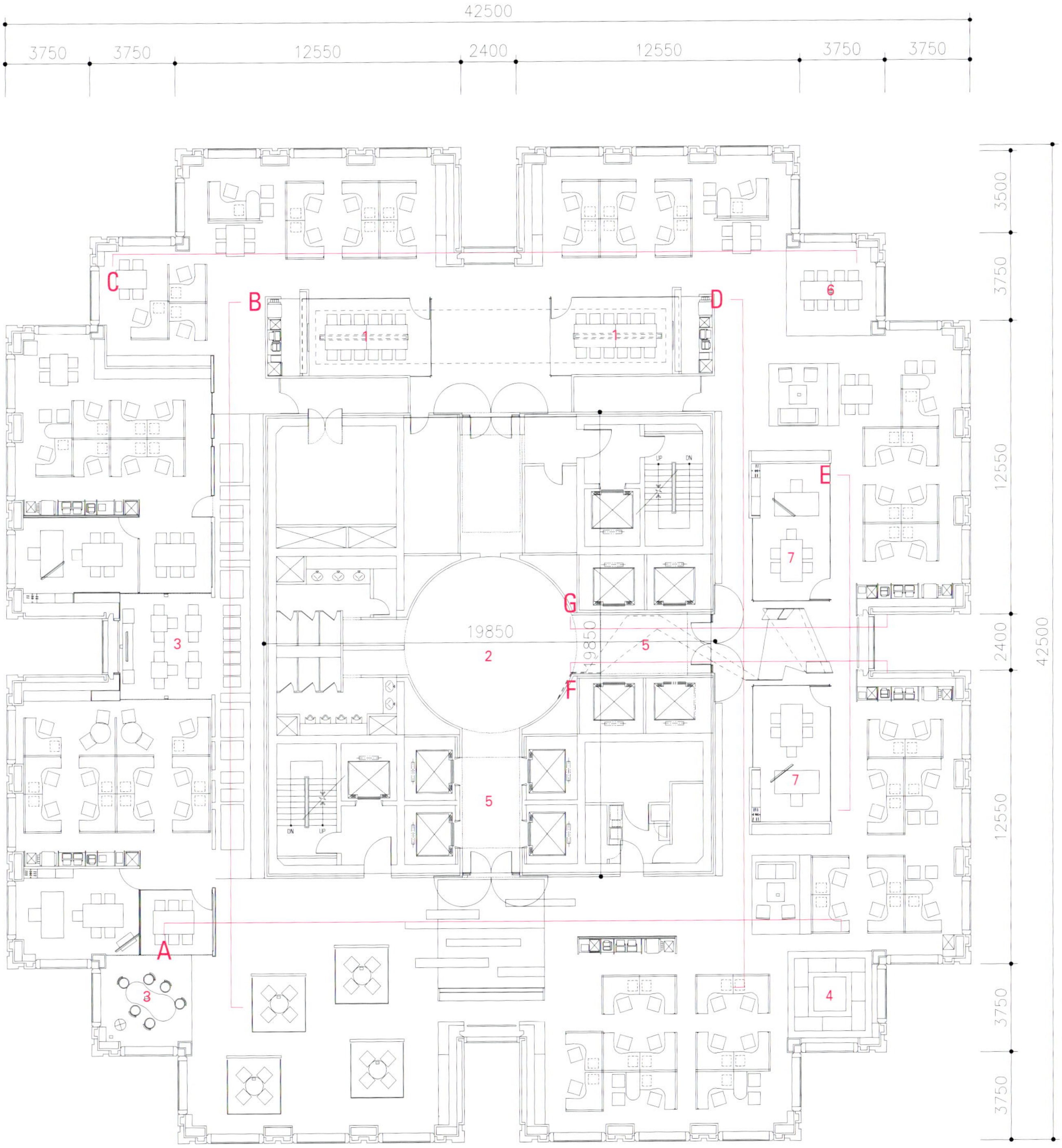

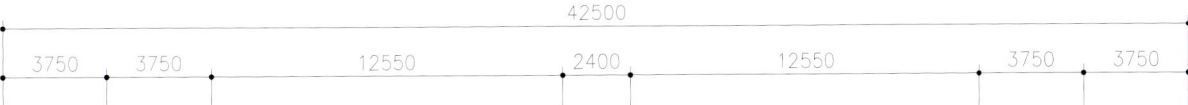

평면도 / floor plan

1 MEETING ROOM **2** CENTER HALL **3** MULTI - CELL **4** MULTI - CELL FOR SEATING **5** ELEVATOR HALL **6** OPEN MULTI - CELL **7** DIRECTOR'S OFFICE

1 회의실 **2** 중앙 홀 **3** 멀티 셀 **4** 좌식 멀티 셀 **5** 엘리베이터 홀 **6** 오픈 멀티 셀 **7** 본부장실

KTB INVESTMENT & SECURITIES HEAD OFFICE | KTB 투자증권 본사 사무공간

MULTI CELL

ART PIGMENT

MEETING AREA

COMMUNICATION WALL

평면 다이어그램 / plan diagram

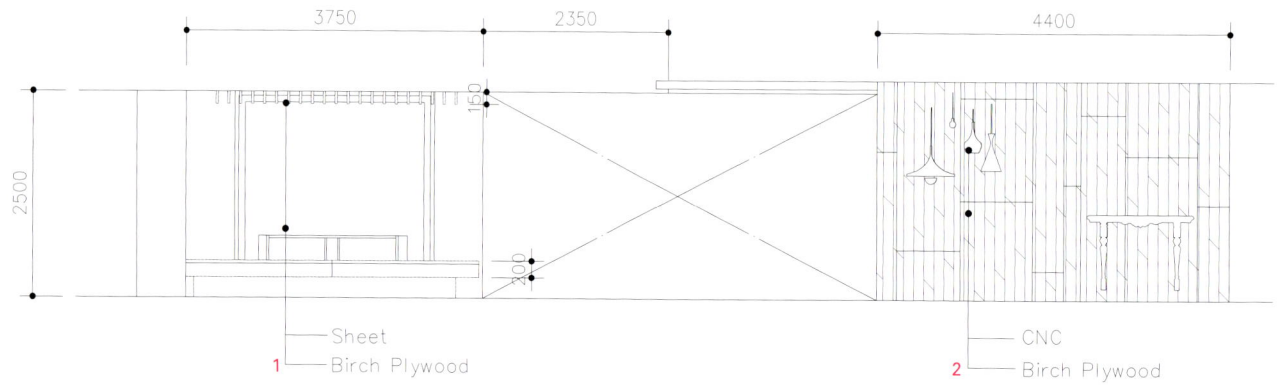

로비 입면 A / lobby elevation A

1 SHEET / BIRCH PLYWOOD 2 CNC / BIRCH PLYWOOD

1 시트 / 자작나무 합판 2 CNC / 자작나무 합판

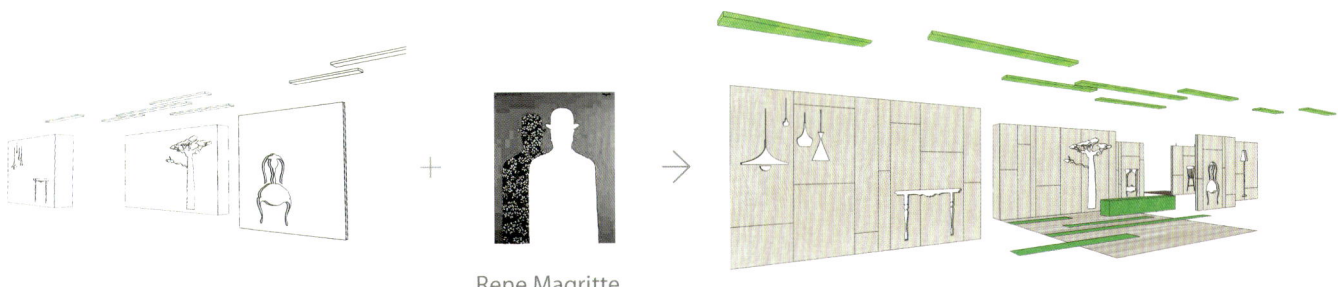

아트 피그먼트 다이어그램 / art pigment diagram

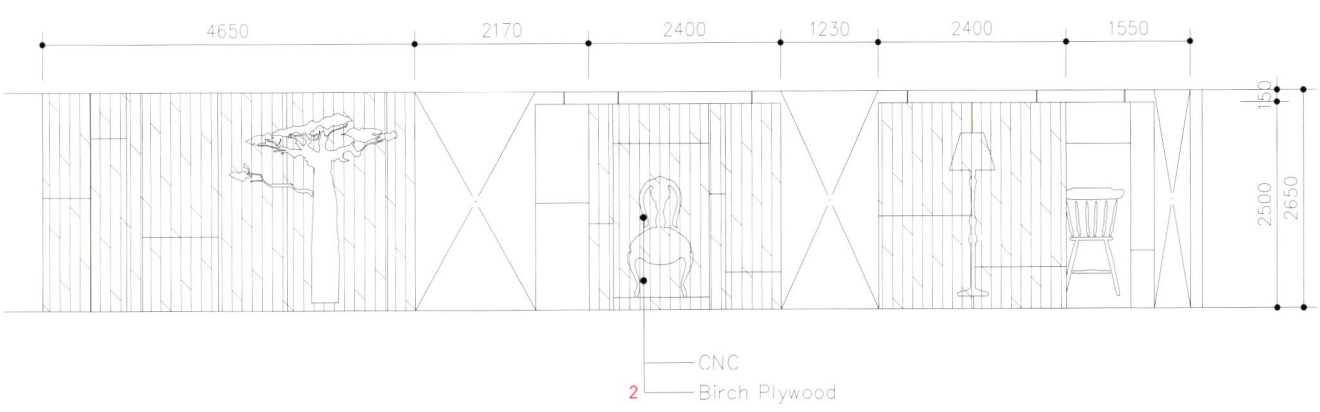

로비 입면 A / lobby elevation A

1 PAINTED 2 PAINTED / WOOD 3 BIRCH PLYWOOD / STONE 4 BIRCH PLYWOOD / WOOD

1 도장 2 도장 / 목재 3 자작나무 합판 / 석재 4 자작나무 합판 / 목재

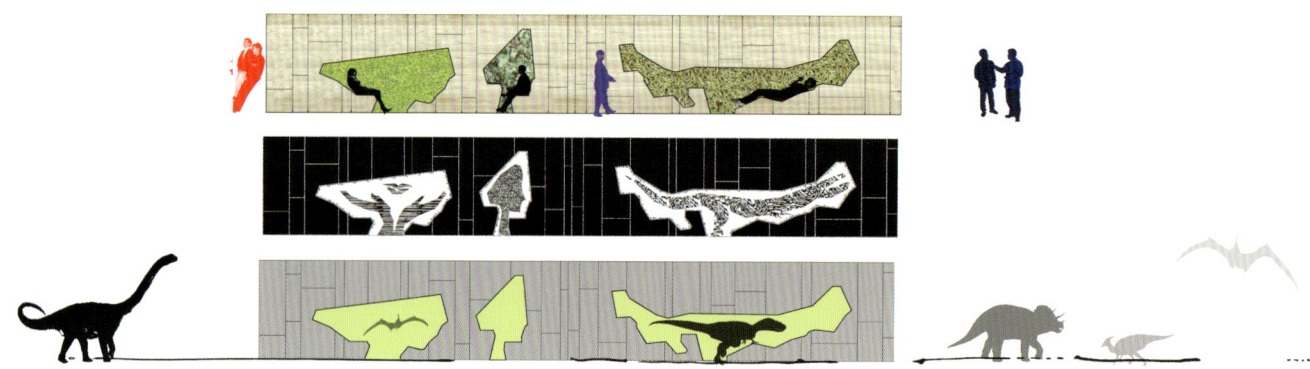

벽체 다이어그램 / wall diagram

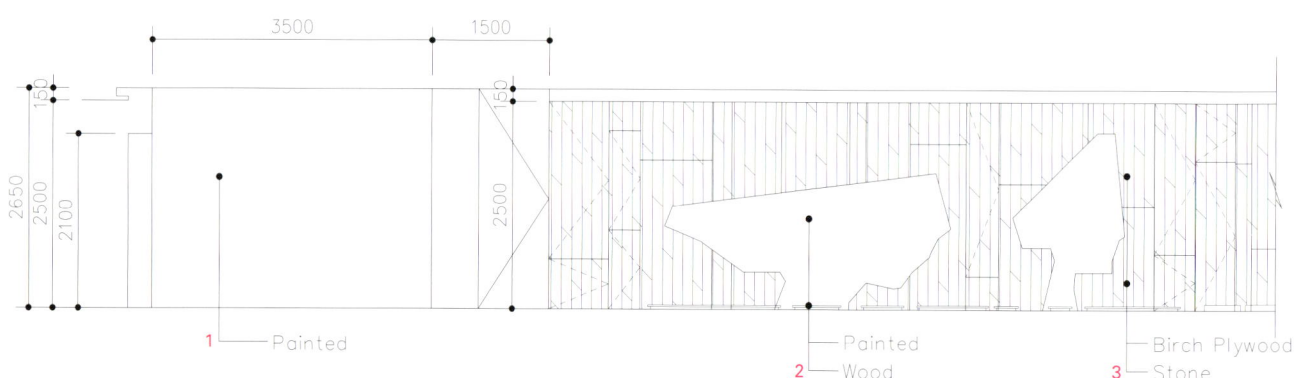

커뮤니케이션 월 입면 B / communication wall elevation B

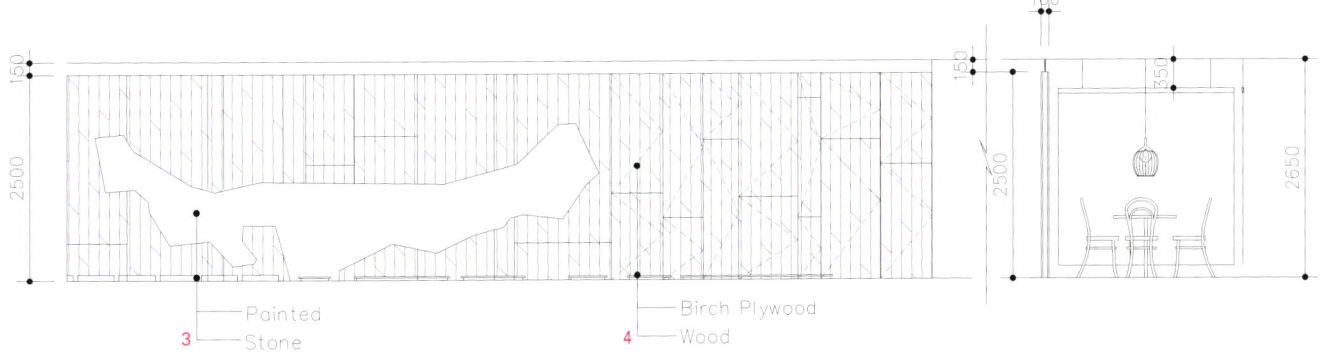

커뮤니케이션 월 입면 B / communication wall elevation B

1 SHEET / BIRCH PLYWOOD 2 BIRCH PLYWOOD 3 SHEET 4 PAINTED

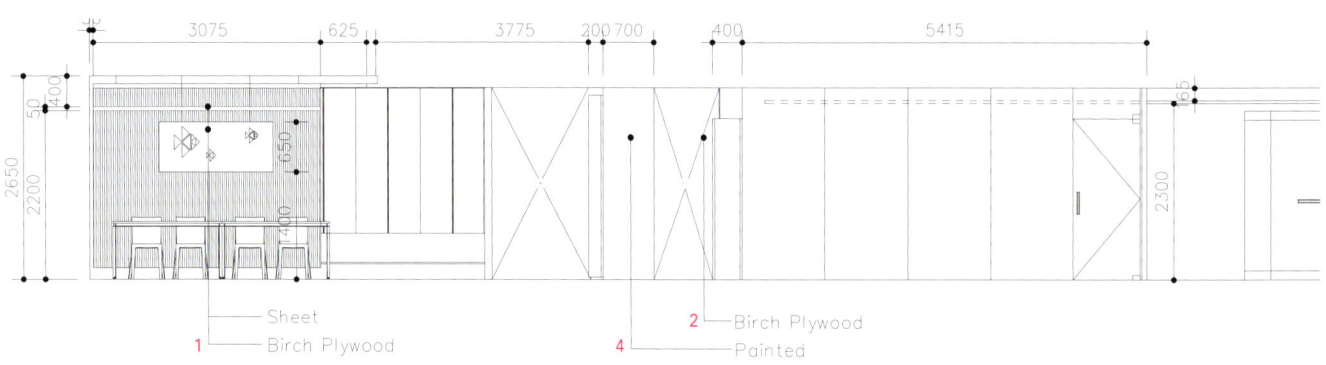

회의실 입면 C / meeting room elevation C

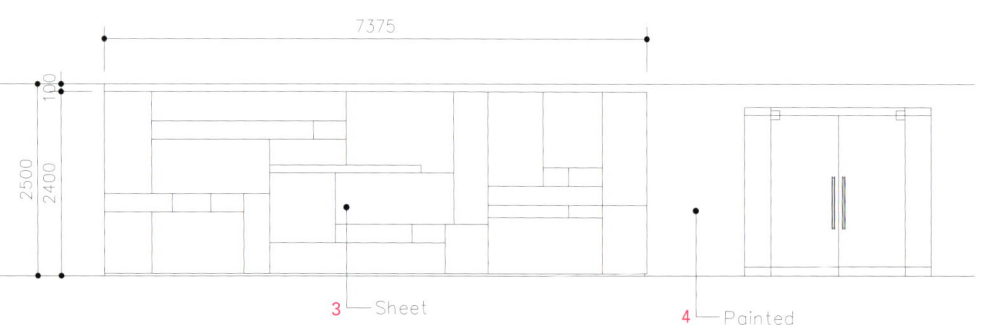

복도 입면 D / corridor elevation D

1 시트 / 자작나무 합판 2 자작나무 합판 3 시트 4 도장

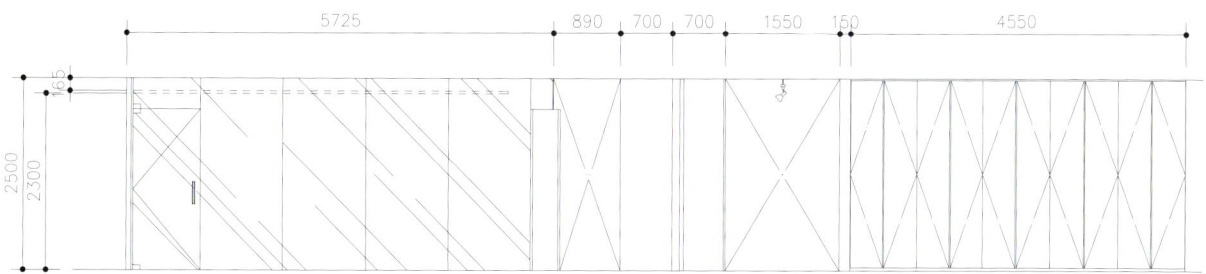

회의실 입면 C / meeting room elevation C

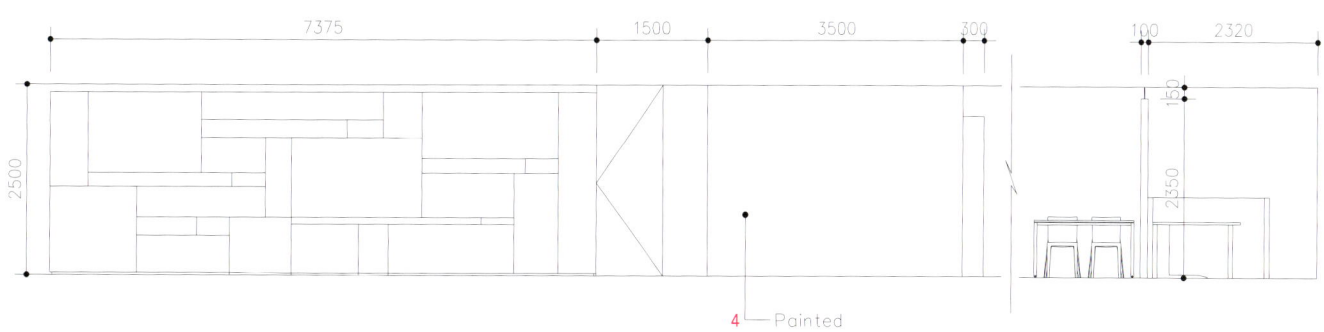

복도 입면 D / corridor elevation D

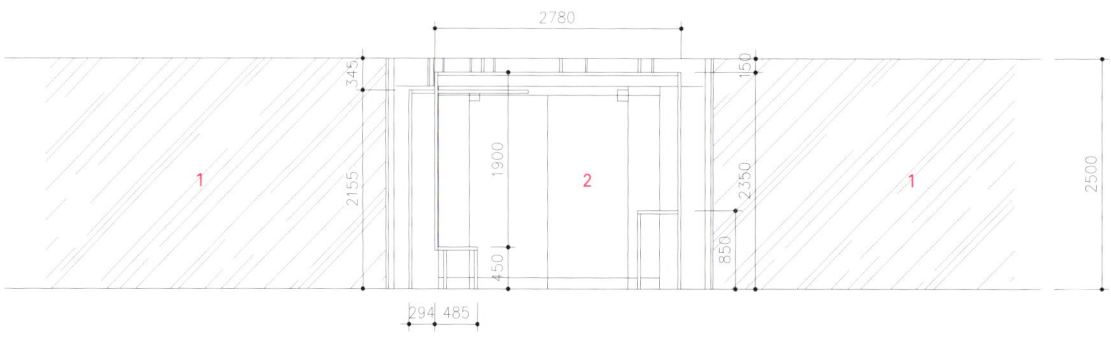

아트 피그먼트 공간 입면 E / art pigment space elevation E

1 DIRECTOR'S OFFICE 2 ART PIGMENT CORRIDOR

1 본부장실 2 아트 피그먼트 코리더

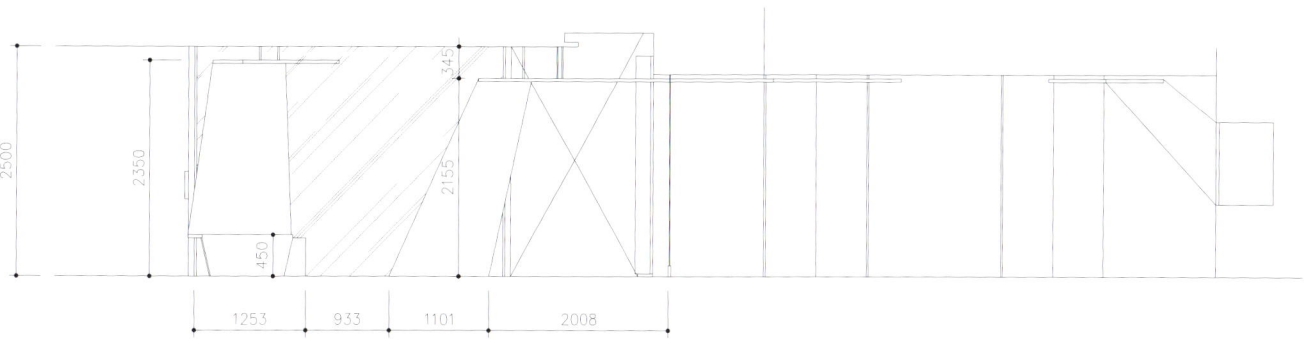

아트 피그먼트 공간 입면 F / art pigment space elevation F

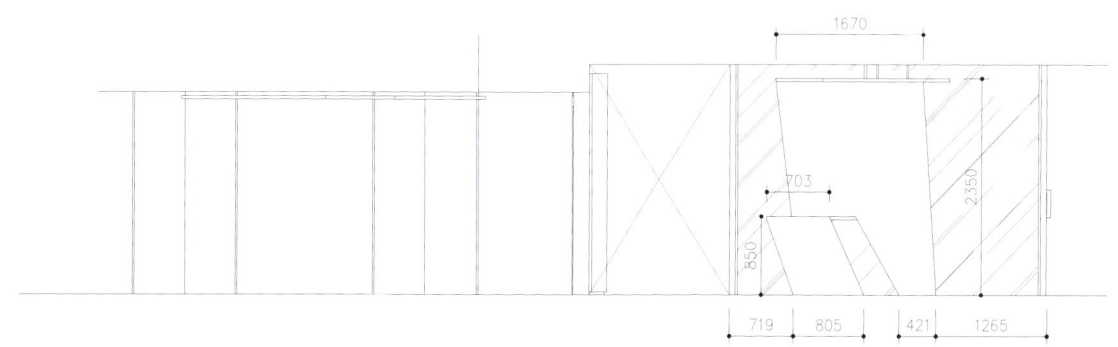

아트 피그먼트 공간 입면 G / art pigment space elevation G

COMING SPRING OF FLOWER

스튜디오 베이스 | 전범진, 원장은, 장병익 studio Vase | Jun Bum Jin, Won Jang Eun, Jang Byung Ik

ㄷ자 형태의 건물 4층에 위치한 '꽃피는 봄이 오면' 은 영화 포스터를 제작하는 회사이다. 건축주는 기존의 공간에서 느꼈던 지루함을 없애고, 봄처럼 화사하게 바꾸자는 제안을 하였다. '달항아리' 를 모티브로 하여 동양의 소박한 미를 담고 있는 달항아리 특유의 부드러움과 자연스러움을 표현하였다. 장식이나 기교가 없는 단순한 형태의 순백색의 담백함이 진솔하고 담담하지만, 열정적인 체취를 채운 공간으로 완성되었다. 주목할 만한 요소는 철체 프레임 책장이다. 한쪽 벽면 전체를 활용한 철제 프레임은 ㄷ자 형태가 갖는 폐쇄성을 극복하고자 나온 결과물이다. 레이어가 겹겹이 포개어진 철제 프레임은 벽의 기능과 파티션, 책장의 기능을 한다. 좁고 단일한 공간 내에 시각적으로 다양한 느낌을 부여하고 개방적인 공간감을 전달한다. 보이드한 프레임은 외부와 소통을 유도하고, 공간 분배의 중심이 된다.

'Coming Spring of Flower' is a movie poster production company. According to their request to change the corny space to a fresh one like spring blossoms, designers designed a ㄷ- shape building inspired by 'Dalhan-gari(nine moon shaped jars of white porcelain)' and attempted to represent the tenderness and natural beauty of Dalhangari that shows a simple beauty of the Oriental arts. The pure white space in a simple shape without decoration and ornament is serene and moderate, but with full of passion of those who work in there. What is remarkable in the 'Coming Spring of Flower' office building which was entirely completed in white from the traditional - motif floor and niche to the ceilings and furniture is the steel-framed library. The steel frame that occupies a whole wall is a result to offset the closeness that the ㄷ- shape building has. The steel frame functions as wall, partition and bookshelf by stacking the layers. It creates various atmosphere within the narrow and simple space visually while making the space look more open and spacious. The void frame becomes the center of the divided spaces by leading communication with the outside of the office.

위치 서울특별시 강남구 논현동 30-6 Mass message 4층
용도 업무
면적 164.3㎡
마감 바닥 – 에폭시, 벽 / 천장 – 비닐 페인트
설계기간 2010. 2 ~ 3
시공 2010. 3
사진 박우진

Location 4F, Mass message, 30-6 Nonhyun-dong, Gangnam-gu, Seoul
Use Office
Area 164.3㎡
Finishing Floor - Epoxy, Wall / Ceiling - Vinyl Paint
Design period 2010. 2 ~ 3
Construction 2010. 3
Photographer Park Woo Jin

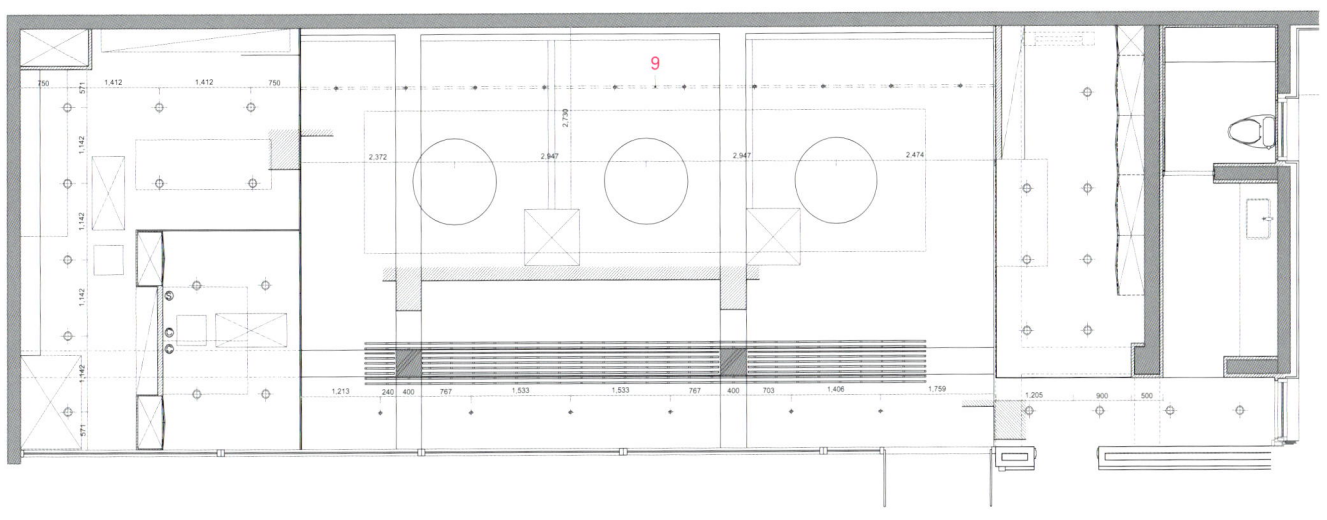

천장도 / ceiling plan

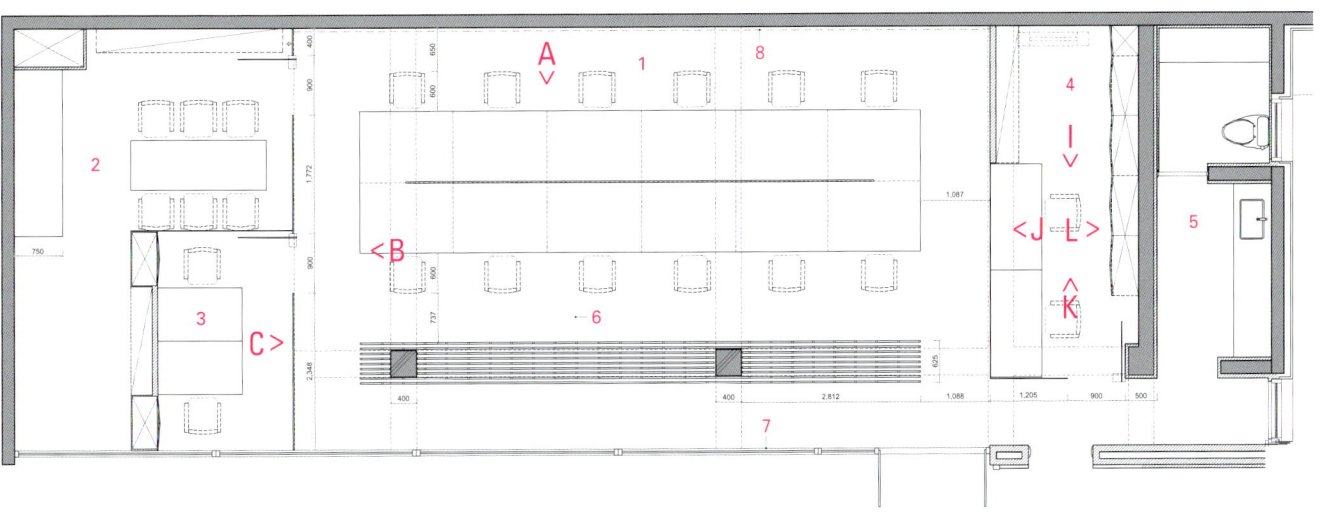

평면도 / floor plan

1 DESIGN ROOM 2 WORK ROOM 3 PLANNING ROOM 4 SUPERVISION ROOM 5 HOT-WATER SERVICE ROOM 6 CLEAR EPOXY 7 CURTAIN 8 GALLERY RAIL 9 TRACK LIGHTING(HALOGEN 10EA)

1 디자인실 2 작업실 3 기획실 4 관리실 5 탕비실 6 투명 에폭시 7 커튼 8 갤러리 레일 9 트랙 조명(할로겐 10개)

COMING SPRING OF FLOWER | 꽃피는 봄이 오면

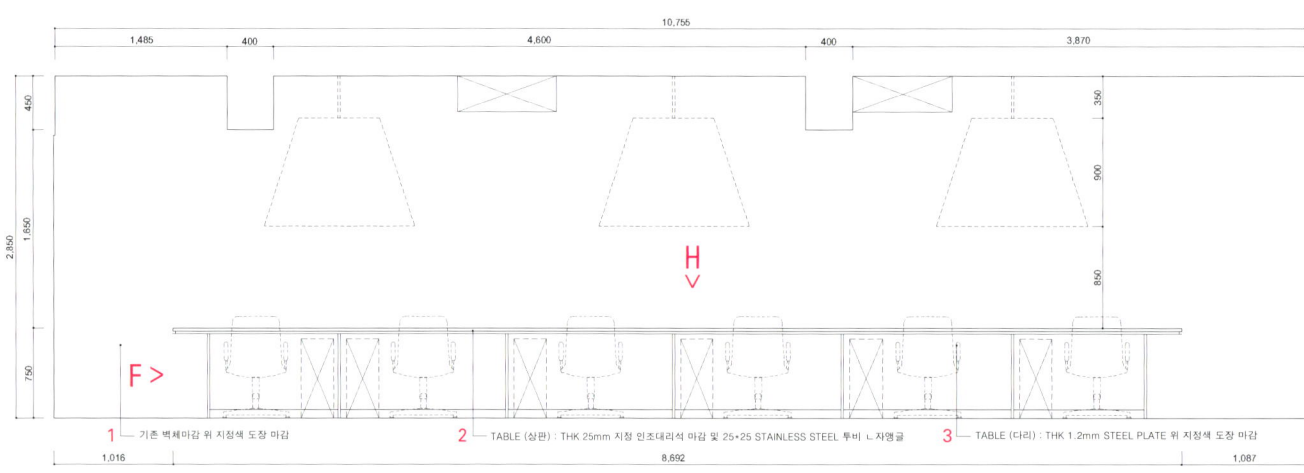

다지인실 입면 A / design room elevation A

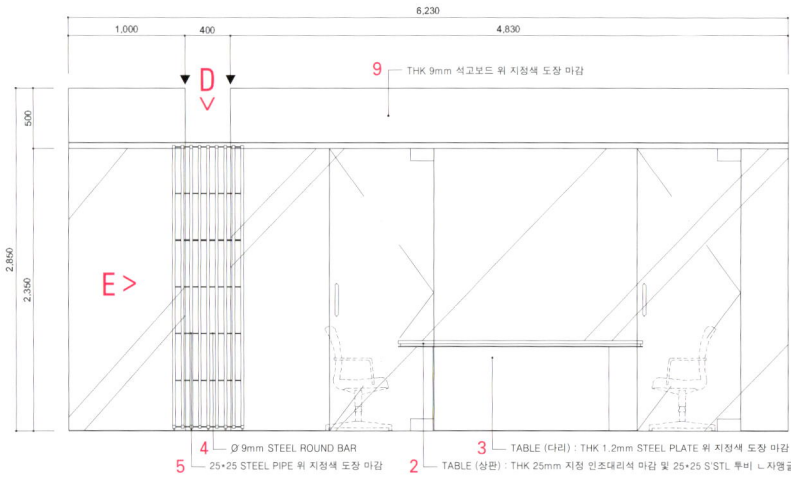

다지인실 입면 B / design room elevation B

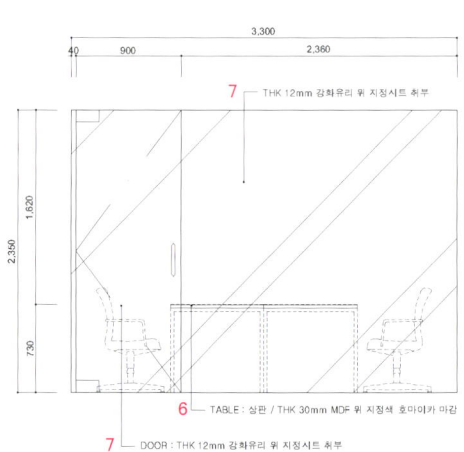

기획실 입면 C / planning room elevation C

1 APPOINTED PAINTING FINISH ON EXISTING WALL FINISH 2 T25 APPOINTED ARTIFICIAL MARBLE FINISH AND L-25X25 STAINLESS STEEL ANLGE 3 T1.2 APPOINTED PAINTING FINISH ON STEEL PLATE 4 Ø9 STEEL ROUND BAR 5 APPOINTED PAINTING FINISH ON 25X25 STEEL PIPE 6 APPOINTED COLOR FORMICA FINISH ON T30 MDF 7 DOOR : APPOINTED SHEET ON T2 TEMPERED GLASS 8 APPOINTED COLOR PAINTING FINISH ON EXISTING COLUMN FINISH

1 기존 벽체 위 지정색 도장 마감 2 T25 지정 인조 대리석 마감 및 25X25 스테인리스 스틸 투비 ㄴ자 앵글 3 T1.2 철제 플레이트 위 지정색 도장 마감 4 Ø9 철제 라운드 바 5 25X25 철제 파이프 위 지정색 도장 마감 6 T30 MDF 위 지정색 호마이카 마감 7 문 : T2 강화 유리 위 지정 시트 취부 8 기존 기둥 위 지정색 도장 마감

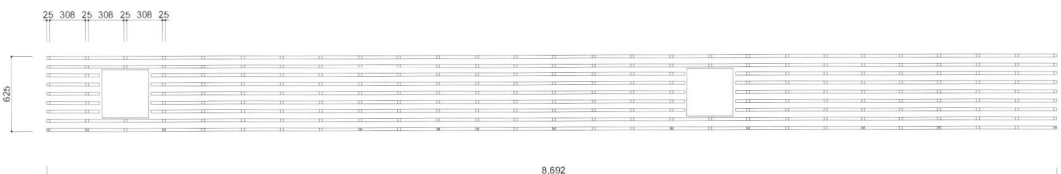

책장 평면 D / bookshelf plan D

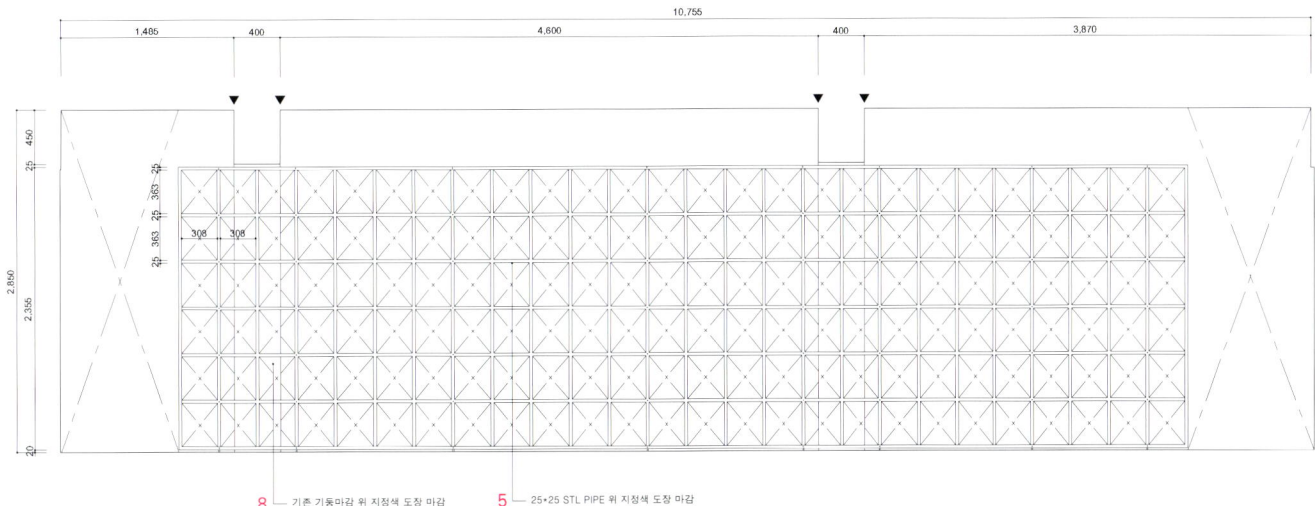

책장 입면 E / bookshelf elevation E

1 FORMICA FINISH ON T18 MDF 2 30X30 STEEL PIPE, APPOINTED COLOR FORMICA FINISH ON T9 MDF ON T9 PLYWOOD 1PLY 3 T12.5 APPOINTED ARTIFICIAL MARBLE 4 T12 PLYWOOD 5 25X40 STEEL PIPE 6 L-25X25 ANGLE 7 T10 APPOINTED SHEET ON TEMPERED GLASS 8 TABLE : APPOINTED FORMICA FINISH ON T30 MDF MDF 9 APPOINTED PAINTING FINISH ON T9 GYPSUM BOARD 10 APPOINTED PAINTING FINISH ON EXISTING WALL FINISH

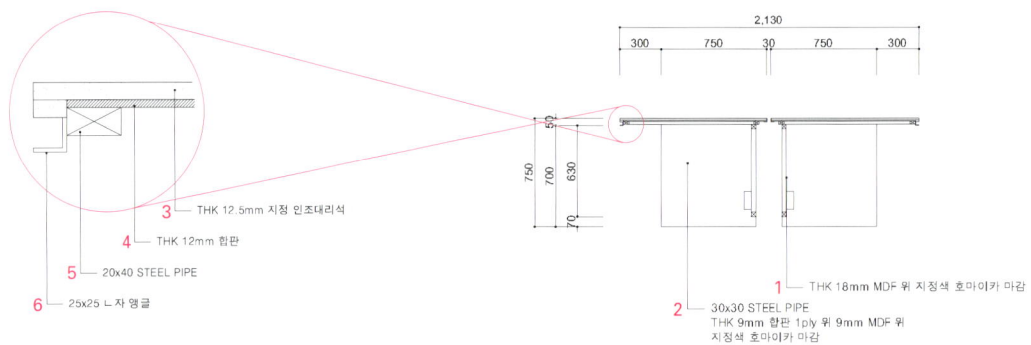

3 — THK 12.5mm 지정 인조대리석
4 — THK 12mm 합판
5 — 20x40 STEEL PIPE
6 — 25x25 ㄴ자 앵글

1 — THK 18mm MDF 위 지정색 호마이카 마감
2 — 30x30 STEEL PIPE
THK 9mm 합판 1ply 위 9mm MDF 위
지정색 호마이카 마감

책상 측면 F / desk side view F

2 — 30x30 STEEL PIPE
THK 9mm 합판 1ply 위 9mm MDF 위
지정색 호마이카 마감

책상 정면 G / desk front view G

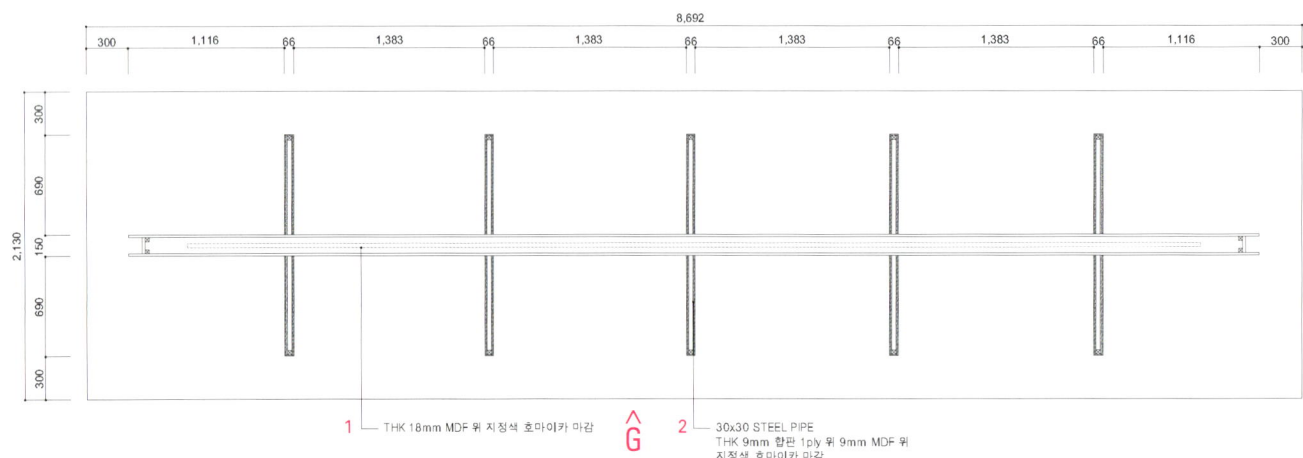

1 — THK 18mm MDF 위 지정색 호마이카 마감
2 — 30x30 STEEL PIPE
THK 9mm 합판 1ply 위 9mm MDF 위
지정색 호마이카 마감

책상 평면 H / desk plan H

1 T18 MDF 위 호마이카 마감 2 30X30 스틸 파이프, T9 합판 1겹 위 T9 MDF 위 지정색 호마이카 마감 3 T12.5 지정 인조대리석 4 T12합판 5 25X40 철제 파이프 6 25X25 ㄴ자 앵글 7 T10 강화 유리 위 지정 시트 취부 8 책상 : T30 MDF 위 지정색 호마이카 마감 9 T9 석고보드 위 지정색 도장 마감 10 기존 벽체 위 지정색 도장 마감

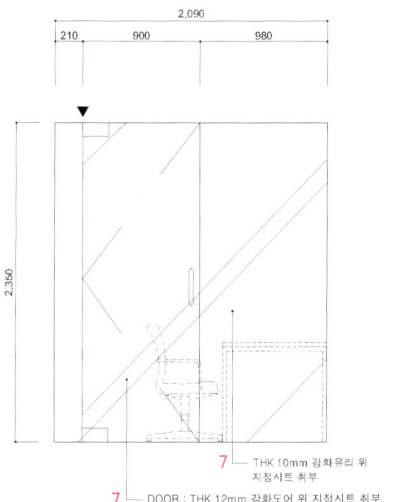

관리실 입면 I / supervision room elevation I

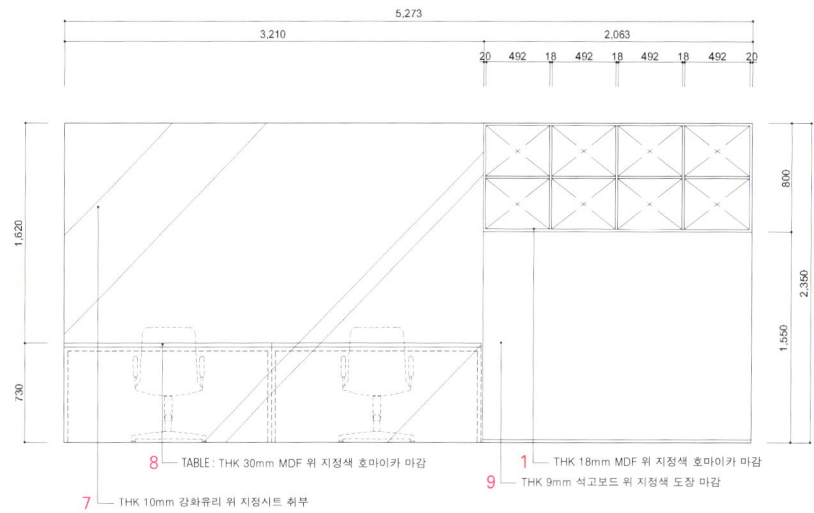

관리실 입면 J / supervision room elevation J

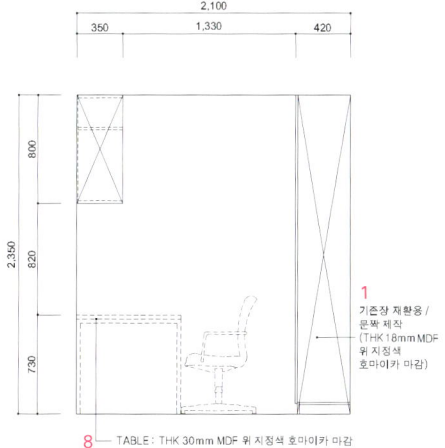

관리실 입면 K / supervision room elevation K

관리실 정면 L / supervision room front view L

ABILITY SYSTEMS

(주)이우진어소시에이트 | 이우진   LEE WOO-JIN ASSOCIATE | Lee Woo Jin

토탈 IT 솔루션 기업인 ABILITY의 아이덴티티를 표현하면서, 젊고 혁신적인 기업 분위기를 살려줄 공간 디자인이 필요했다. 전체적으로 과장되지 않은 절제된 아름다움을 추구하며 진보적이고 창의적인 공간을 디자인 방향으로 잡았고, IT의 미래적인 이미지를 표현하고자 했다. 입구에 들어서면 에폭시 처리된 매끈한 바닥면에서 오는 세련되고 미래적인 이미지를 느낄 수 있다. 원목의 의자와 테이블을 배치해 딱딱하고 차가운 느낌을 완화시키고 공간의 긴장감을 풀어주었다. 공간을 가로지르는 큰 획을 유연하고 매끈한 곡선으로 풀어 마치 유선형의 기체에 들어온 듯한 느낌이 들도록 했다. 특히 파사드는 실제 우주선의 입구와 같은 사이버 공간을 연상시킨다. 곡선의 벽 위로 발광하는 간접 조명을 사용하여 부유하고 있는 듯한 느낌을 더하고, 유영하는 곡선의 느낌은 창과 문의 둥근 사다리 모양 실루엣으로 연결된다. 특히 원이라는 단순한 기하학 패턴으로 크기를 변주하고 반복, 적용하여 공간을 부드럽게 풀어주었다. 둥근 의자와 전등에서도 곡선의 흐름이 반복되어 공간에 활력을 불어넣는 요소가 된다.

Young and innovative design was required to represent the identity of total IT - solution company 'ABILITY' in building their office. In general, we seek moderate and controlled beauty and focused on designing a progressive and creative space while highlighting the future - going images of the IT company. The first thing that comes up when entering into the entrance is the futuristically - finished sleek and sophisticated epoxy floor in white and grey color. The warm wooden colored chairs and table set reduces intensity of the space by offsetting the cold and stiff mood. The outstanding line that penetrates the whole space is transformed into a flexible and smooth curve which makes us feel as if we are in an oval aircraft, and especially facade reminds us of a cyber space like entrance of a spaceship. The indirect lighting over the curvy wall looks like floating and the floating curves are related to the trapezoid silhouette created by the windows and the door. The simple and geometric pattern of circle adds vibrancy to the space through its repetition, as well as the round chairs and lightings in which the circle pattern is repeated.

위치 서울특별시 성동구 성수동 308-4
용도 업무
면적 419.80m²
마감 바닥 - 화이트 우레탄 위 투명 에폭시, 카펫 / 벽 - 석고 보드 위 백색 래커 도장, 시트 / 천장 - 석고 보드 위 비닐 페인트 도장
설계기간 2010. 6 ~ 7
공사기간 2010. 7 ~ 8
디자인팀 석희옥, 도은정
시공팀 전민관, 전재경
사진 (주)이우진어소시에이트 제공

Location 308-4, Seongsu-dong, Seongdong-gu, Seoul
Use Office
Area 419.80m²
Finishing Floor - White urethane on transparent epoxy, Carpet / Wall - White lacquer painting on gypsum board, Sheet / Ceiling - Vinyl paint painting on gypsum board
Design period 2010. 6 ~ 7
Construction period 2010. 7 ~ 8
Photos offer LEE WOO-JIN ASSOCIATE

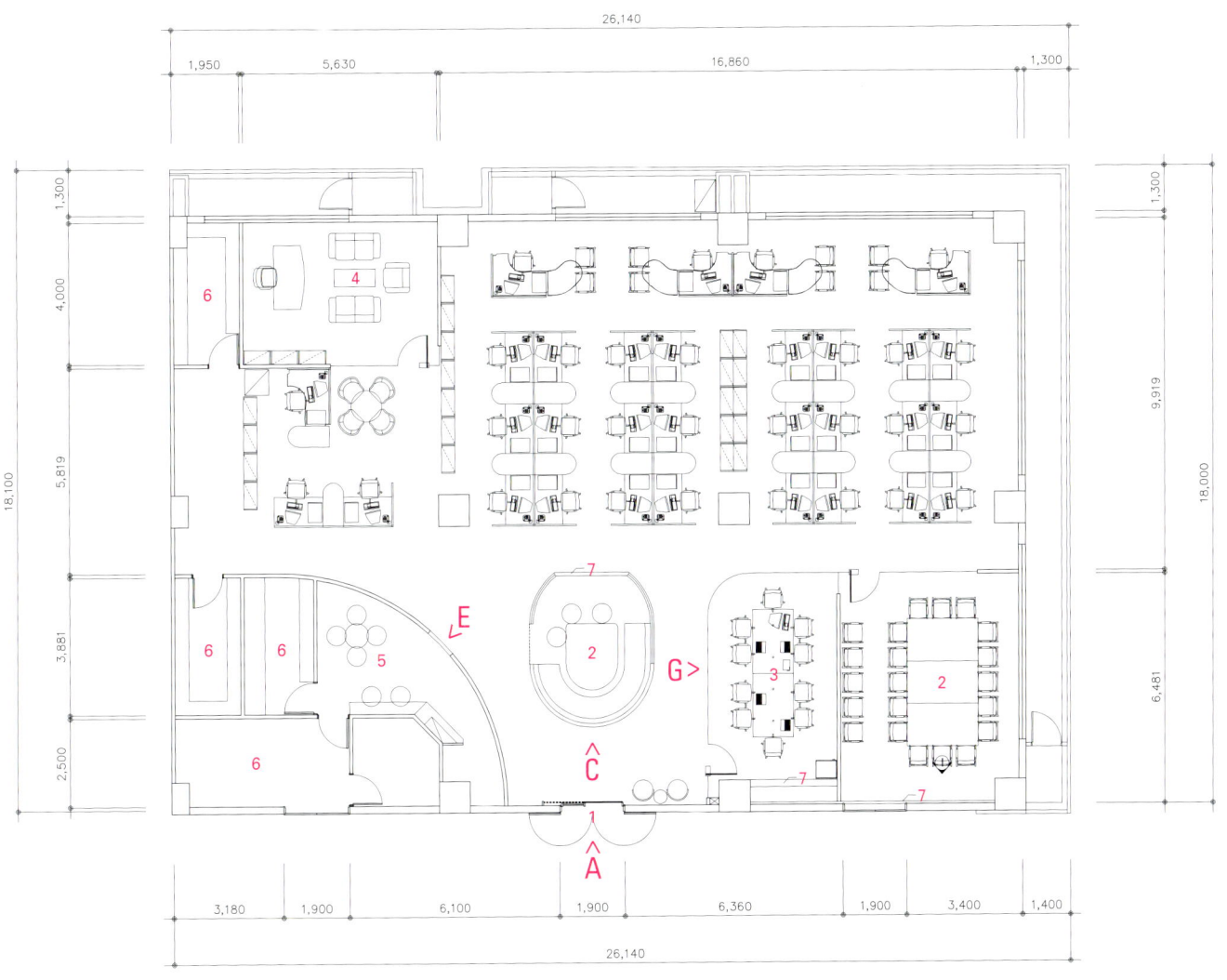

평면도 / floor plan

1 ENTRANCE 2 CONFERENCE ROOM 3 EDUCATION ROOM 4 PRESIDENT ROOM 5 RESTING SPACE 6 STORAGE 7 GLASS BOARD

1 입구 2 회의실 3 교육실 4 회장실 5 휴게실 6 창고 7 유리 보드

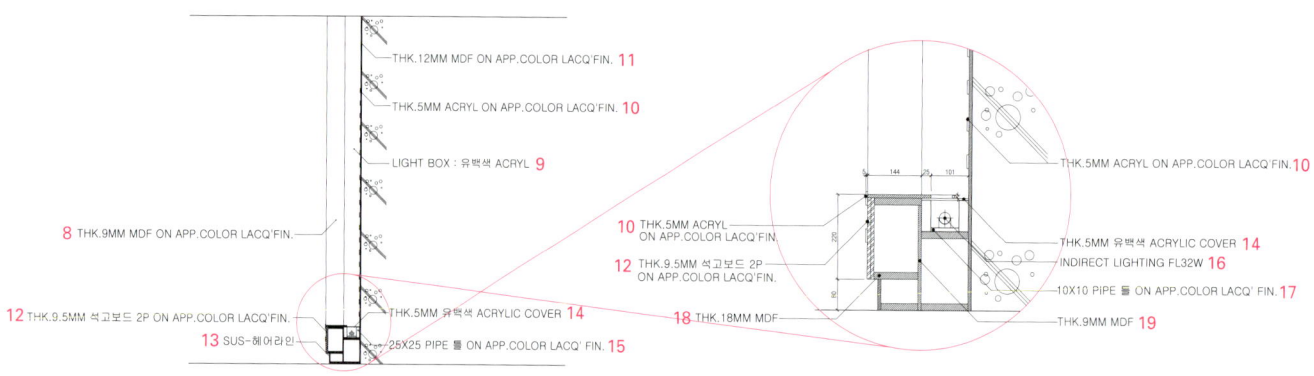

입구 단면 B / entrance section B

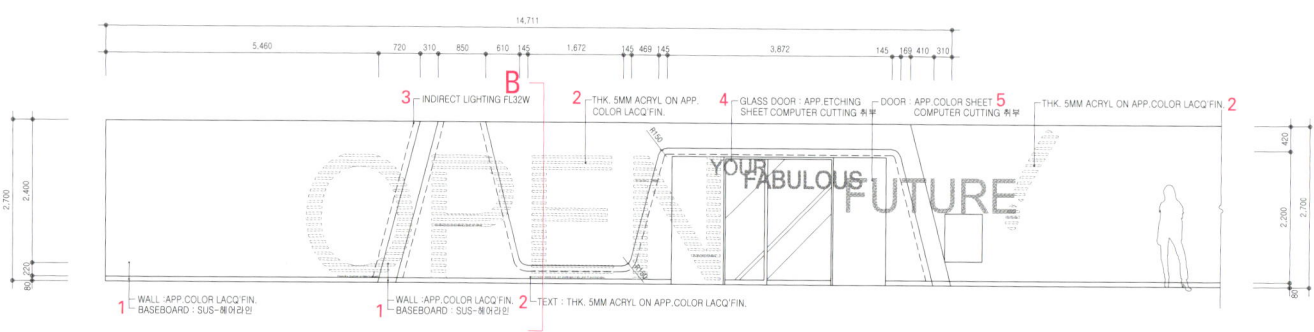

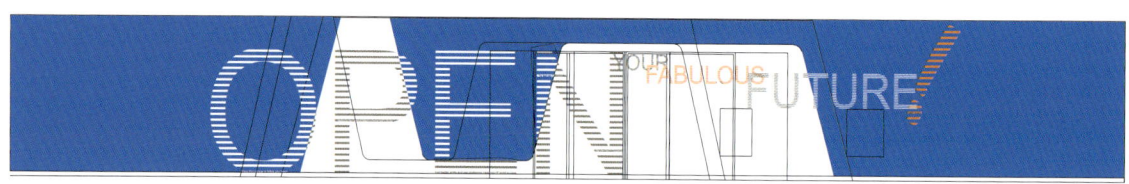

입구 입면 A / entrance elevation A

1 WALL : APP. COLOR LACQ' FIN / BASEBOARD : SUS - HAIRLINE 2 TEXT : T5 ACRYL ON APP. COLOR LACQ. FIN. 3 INDIRECT LIGHTING FL32W 4 GLASS DOOR : APP. ETCHING SHEET COMPUTER CUTTING INSTALLING 5 DOOR : APP. COLOR SHEET COMPUTER CUTTING INSTALLING 6 GLASS WALL : APP. COLOR SHEET COMPUTER CUTTING INSTALLING / GLASS WALL : APP. ETCHING SHEET COMPUTER CUTTING INSTALLING 7 T10 CLEAR GLASS 8 T9 MDF ON APP. COLOR LACQ' FIN. 9 LIGHT BOX : IVORY WHITE ACRYL 10 T5 ACRYL ON APP. COLOR LACQ' FIN. 11 T12 MDF ON APP. COLOR LACQ' FIN. 12 T9.5 GYPSUM BOARD 2P ON APP. COLOR LACQ' FIN. 13 SUS - HAIRLINE 14 T5 IVORY WHITE ACRYLIC COVER 15 25X25 PIPE FRAME ON APP. COLOR LACQ' FIN. 16 INDIRECT LIGHTING FL32W 17 10X10 PIPE FRAME ON APP. COLOR LACQ' FIN. 18 T18 MDF 19 T9 MDF 20 T18 MDF ON APP. COLOR LACQ' FIN. 21 T3 PLYWOOD 3PLY ON APP. COLOR LACQ' FIN. 22 INDIRECT LIGHTING DECO 28W(T5) 23 WHITE VINYL PAINT PAINTING ON T1.5 METAL(GALVA) LIGHTING BOX 24 20X20 PIPE 25 SOFA(MAKING) : APP. WHITE LEATHER ON STITCH FIN.

1 벽 : 지정 컬러 래커 마감 / 걸레받이 : SUS – 헤어라인 2 텍스트 : 지정 컬러 래커 마감 위 T5 아크릴 3 간접 조명 FL32W 4 유리 도어 : 지정 에칭 시트 컴퓨터 커팅 취부 5 도어 : 지정 컬러 시트 컴퓨터 커팅 취부 6 유리 벽 : 지정 컬러 시트 컴퓨터 커팅 취부 / 유리 벽 : 지정 에칭 시트 컴퓨터 커팅 취부 7 T10 투명 유리 8 지정 컬러 래커 마감 위 T9 MDF 9 조명 박스 : 유백색 아크릴 10 지정 컬러 래커 마감 위 T5 아크릴 11 지정 컬러 래커 마감 위 T12 MDF 12 지정 컬러 래커 마감 위 T9.5 석고보드 2겹 13 SUS – 헤어라인 14 T5 유백색 아크릴 커버 15 지정 컬러 래커 마감 위 25X25 파이프 틀 16 간접 조명 FL32W 17 지정 컬러 래커 마감 위 10X10 파이프 틀 18 T18 MDF 19 T9 MDF 20 지정 컬러 래커 마감 위 T18 MDF 21 지정 컬러 래커 마감 위 T3 합판 3겹 22 간접 조명 데코 28W(T5) 23 T1.5 금속(갈바) 라이팅 박스 위 흰색 비닐 페인트 도장 24 20X20 파이프 25 소파(제작) : 스티치 마감 위 지정 흰색 가죽

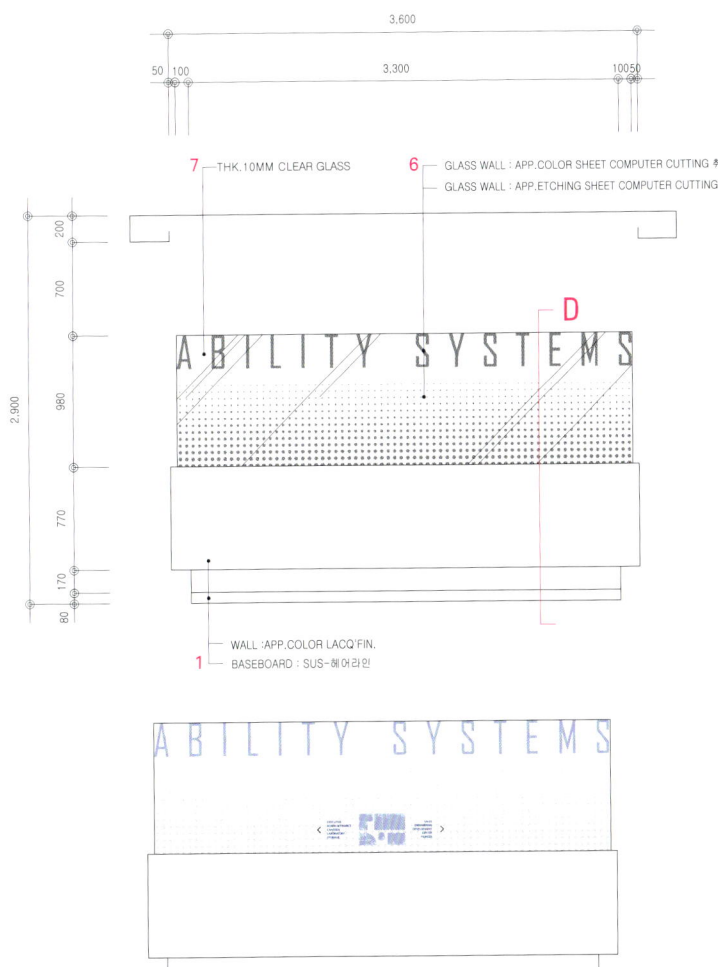

회의실 입면 C / conference room elevation C

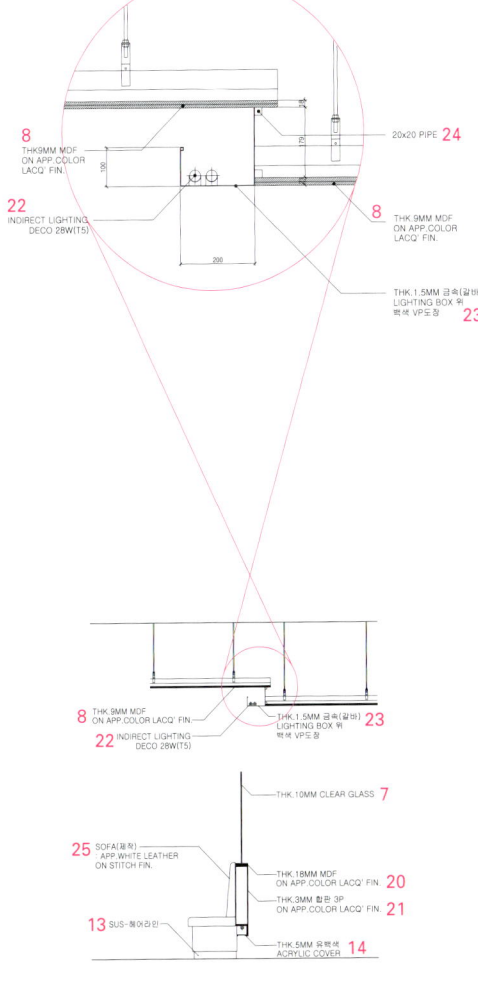

회의실 단면 D / conference room section D

ABILITY SYSTEMS | 어빌리티 시스템즈

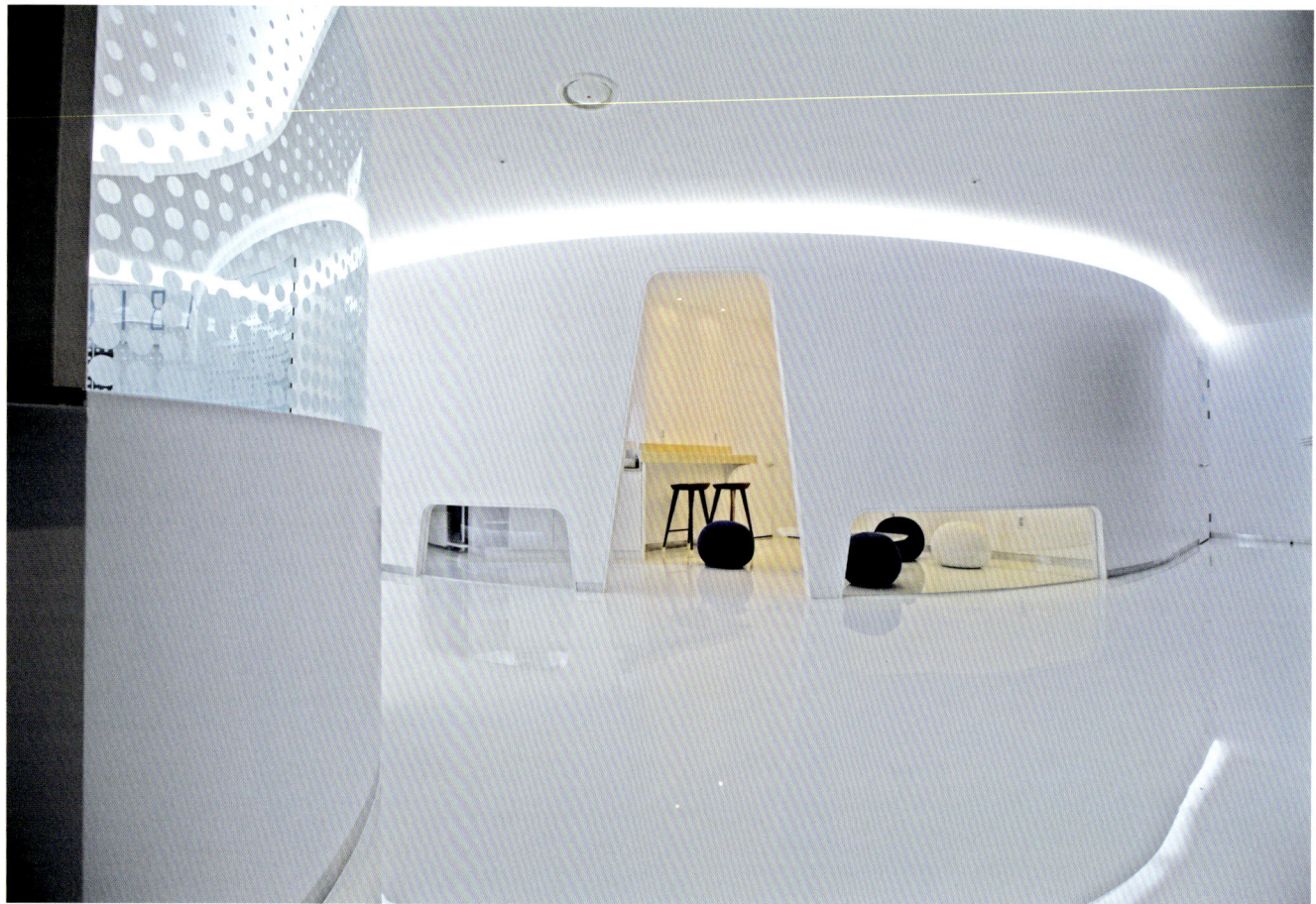

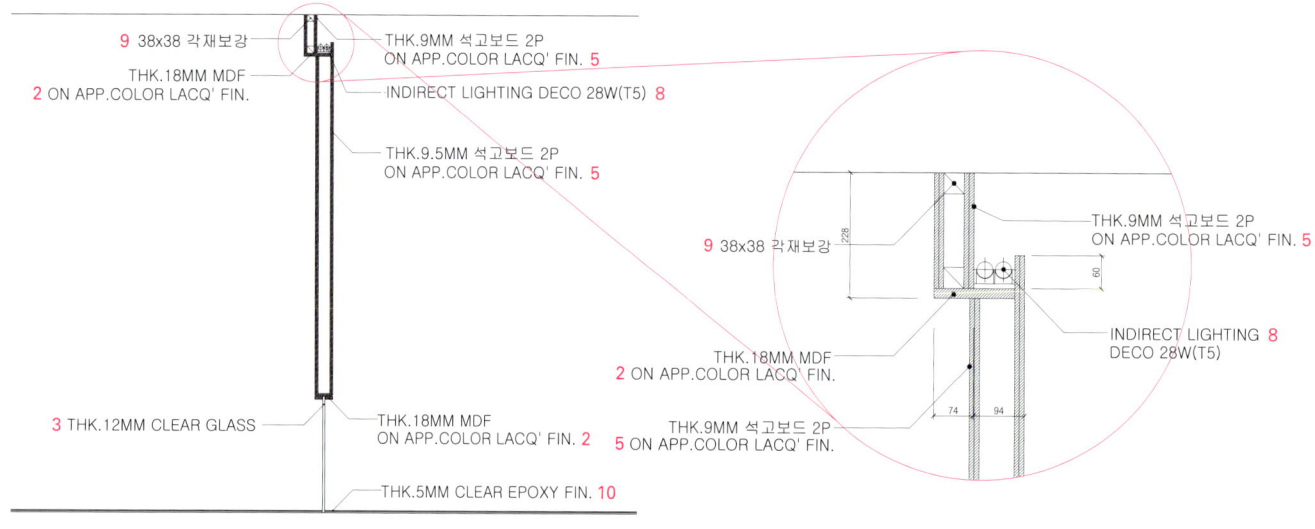

휴게실 단면 F / resting space section F

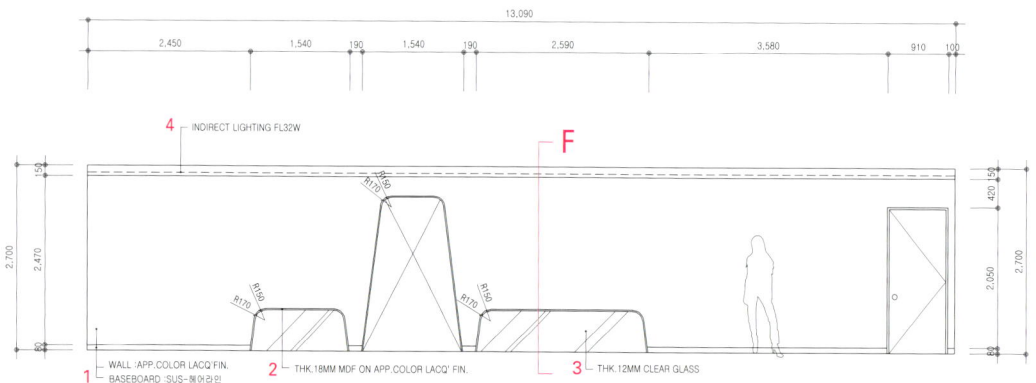

휴게실 입면 E / resting space elevation E

1 WALL : APP. COLOR LACQ' FIN. / BASEBOARD : SUS - HAIRLINE 2 T18 MDF ON APP. COLOR LACQ' FIN. 3 T12 CLEAR GLASS 4 INDIRECT LIGHTING FL32W 5 T9 GYPSUM BOARD 2P ON APP. COLOR LACQ' FIN. 6 BUILT - IN SHELF / BASEBOARD : SUS - HAIRLINE 7 T5 CLEAR EPOXY FIN. 8 INDIRECT LIGHTING DECO 28W(T5) 9 38X38 SQUARE WOOD REINFORCING 10 T5 CLEAR EPOXY FIN. 11 T9 MDF ON APP. COLOR LACQ' FIN. 12 T10 TEMPERED FROST GLASS 13 T15 MDF REINFORCING 14 SUS - HAIRLINE

1 벽 : 지정 컬러 래커 마감 / 걸레받이 : SUS – 헤어라인 2 지정 컬러 래커 마감 위 T18 MDF 3 T12 투명 유리 4 간접 조명 FL32W 5 지정 컬러 래커 마감 위 T9 석고보드 2겹 6 붙박이 선반 / 걸레받이 : SUS – 헤어라인 7 T5 투명 에폭시 마감 8 간접 조명 데코 28W(T5) 9 38X38 각재 보강 10 T5 투명 에폭시 마감 11 지정 컬러 래커 마감 위 T9 MDF 12 T10 강화 프로스트 유리 13 T15 MDF 보강 14 SUS – 헤어라인

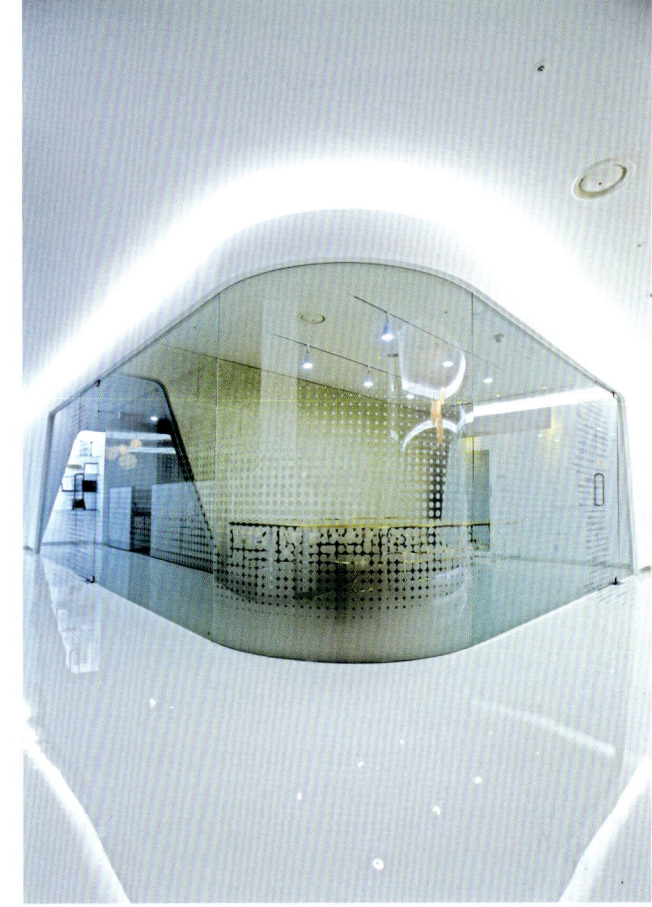

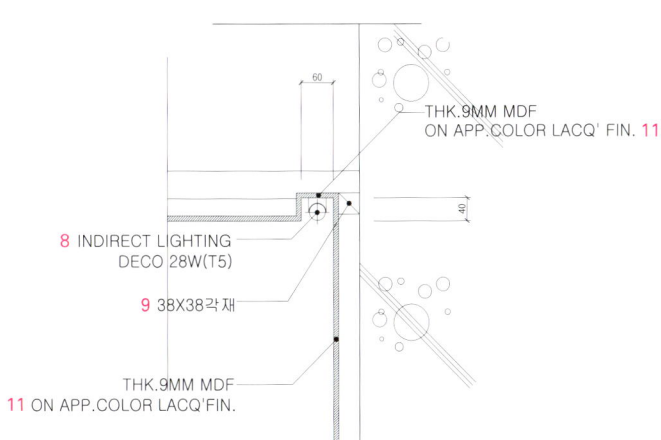

상세 I / detail I

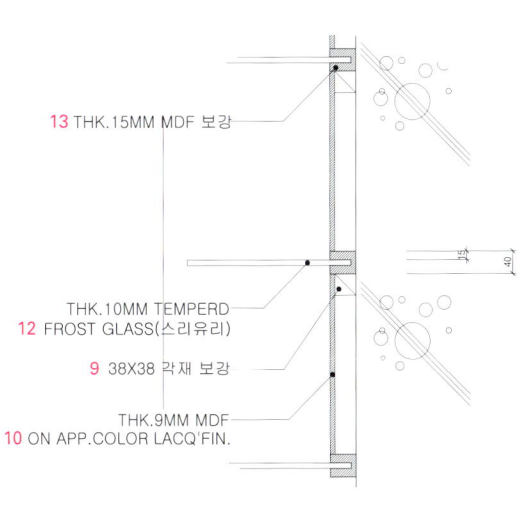

상세 J / detail J

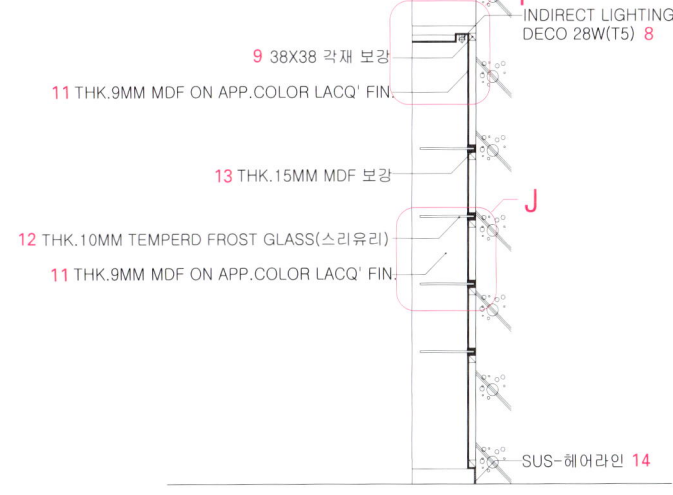

교육실 단면 H / education room section H

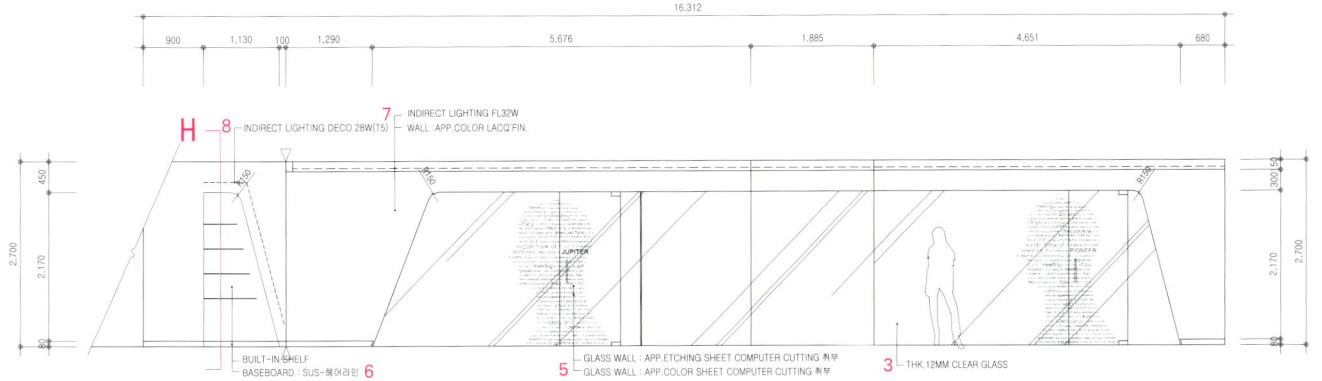

교육실 입면 G / education room elevation G

CROSS-STITCH

(주)이우진어소시에이트 | 이우진   LEE WOO-JIN ASSOCIATE | Lee Woo Jin

(주)유니트 아이엔씨는 미국의 백화점들을 상대로 패션 의류를 생산하는 OEM(Original Equipment Manufacturing) 회사로 출발해 최근에는 직접 상품을 디자인하고 디렉팅하는 ODM(Original Design Manufacturing) 회사이다. 디자인을 가시화시키는 작업을 하는 제조업체로써 자체적인 디자인 회의 또한 많아 공간을 디자인하는데 있어 소통을 이룰 수 있는 공간이 중요한 부분이었다. '선(線)'은 공간 전체를 아우르고 포용하는 요소이다. 선에 의해 연결, 분절되는 공간으로 소통을 가시화하고자 했다. 또한 선은 평평한 바닥을 따라 벽으로 이어져 조명과 가구의 입면으로서의 기능을 하고, 소파와 스툴에 스티치(Stitch)로 새겨지며 그 연속성을 더해간다. 벽을 따라 역동하는 선은 천장에 이르러 트랙 조명 라인과 노출 천장 속에 얽혀 있는 배관 등의 구조물과도 어우러져 입체적으로 표현되었다.

UNIT INC CO.,LTD., started as a OEM(Original Equipment Manufacturing) apparel production company exporting to major department stores in the States, designs and directs apparel ODM(Original Design Manufacturing) procedures. For the company that materializes the design through fashion, communication such as design meeting was the most considerable factor in designing the space. Lines are the core element that embraces and harmonizes the whole space as the space is connected or fragmented by lines. In addition, lines function as elevation for the lightings and furniture, flowing along the floor and the walls. The stitch on the sofa and stool takes the continuity of the lines. The lines are dynamically connected from the walls to the ceilings in which the wires for the lightings on the track meet the lines.

위치 서울특별시 구로구 구로동 197-5
용도 쇼룸, 업무
면적 288.34㎡
마감 바닥 - 셀프 레벨링 위 에폭시 수지 / 벽 - 석고보드 위 컬러 래커 도장, 시트 / 천장 - 석고보드 위 비닐 페인트 도장, 노출콘크리트 위 비닐 페인트 도장
설계기간 2010. 5 ~ 6
공사기간 2010. 6 ~ 7
디자인팀 석희옥, 도은정
시공팀 전민관, 전재경
사진 (주)이우진어소시에이트 제공

Location 197-5, Guro-dong, Guro-gu, Seoul
Use Showroom, Office
Area 288.34m²
Finishing Floor - Epoxy resin on self-leveling / Wall - Color lacquer painting on gypsum board, Sheet / Ceiling - Vinyl paint painting on gypsum board, Vinyl paint painting on exposed concrete
Design period 2010. 5 ~ 6
Construction period 2010. 6 ~ 7
Photos offer LEE WOO-JIN ASSOCIATE

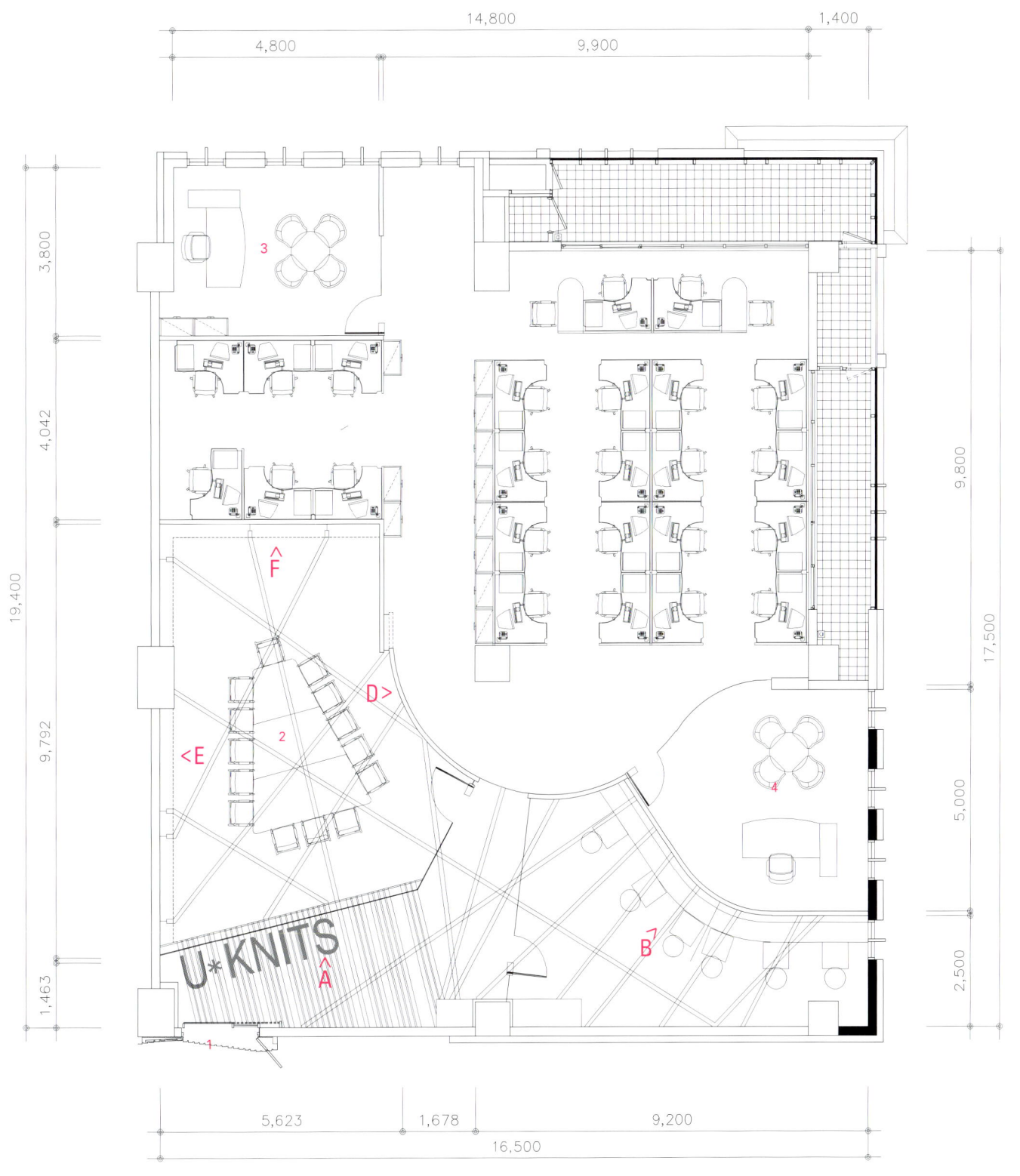

평면도 / floor plan

1 ENTRANCE 2 CONFERENCE ROOM 3 PRESIDENT ROOM 4 DIRECTOR ROOM

1 입구 2 회의실 3 회장실 4 임원실

CROSS-STITCH | 크로스—스티치

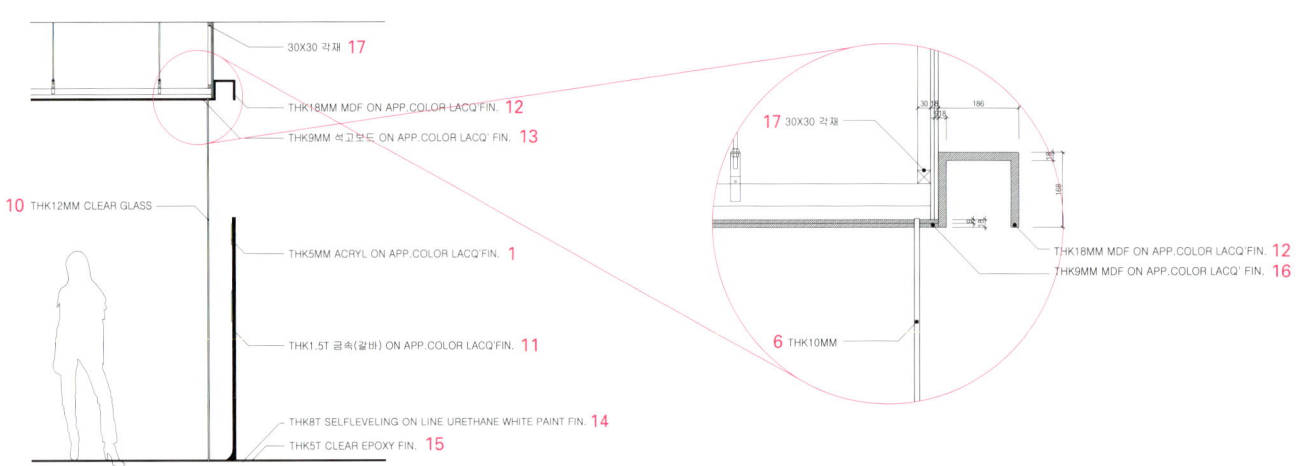

내벽 상세 C / wall detail C

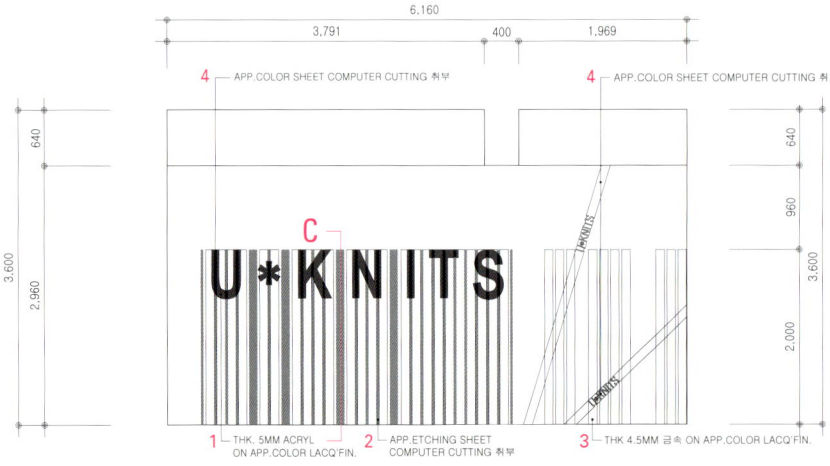

입구 홀 입면 A / entrance hall elevation A

1 T5 ACRYL ON APP. COLOR LACQ' FIN. 2 APP. ETCHING SHEET COMPUTER CUTTING INSTALLING 3 T4.5 METAL ON APP. COLOR LACQ' FIN. 4 APP. COLOR SHEET COMPUTER CUTTING INSTALLING 5 T18 MDF 9MM JOINT ON APP. COLOR LACQ' FIN. 6 T10 CLEAR GLASS 7 INDIRECT LIGHTING FL32W 8 TRACK LIGHTING 9 SOFA(MAKING) : APP. WHITE LEATHER ON STITCH FIN. / T18 MDF 9MM JOINT ON APP. COLOR LACQ' FIN. 10 T12 CLEAR GLASS 11 T5 ACRYL ON APP. COLOR LACQ' FIN. 12 T18 MDF ON APP. COLOR LACQ' FIN. 13 T9 GYPSUM BOARD ON APP. COLOR LACQ' FIN. 14 T8 SELF - LEVELING ON LINE URETHANE WHITE PAINT FIN. 15 T5 CLEAR EPOXY FIN. 16 T9 MDF ON APP. COLOR LACQ' FIN. 17 30X30 SQUARE WOOD

1 지정 컬러 래커 마감 위 T5 아크릴 2 지정 에칭 시트 컴퓨터 커팅 취부 3 지정 컬러 래커 마감 위 T4.5 금속 4 지정 컬러 시트 컴퓨터 커팅 취부 5 지정 컬러 래커 마감 위 T18 MDF 9MM 매지 6 T10 투명 유리 7 간접 조명 FL32W 8 트랙 조명 9 소파(제작): 스티치 마감 위 지정 흰색 가죽 / 지정 컬러 래커 마감 위 T18 MDF 9MM 매지 10 T12 투명 유리 11 지정 컬러 래커 마감 위 T5 아크릴 12 지정 컬러 래커 마감 위 T18 MDF 13 지정 컬러 래커 마감 위 T9 석고보드 14 우레탄 흰색 페인트 마감 위 T8 셀프 – 레벨링 15 T5 투명 에폭시 마감 16 지정 컬러 래커 마감 위 T9 MDF 17 30X30 각재

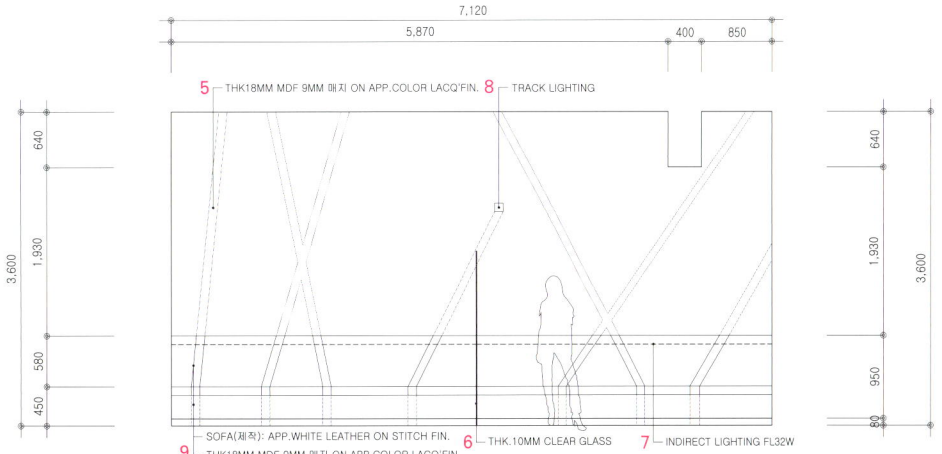

내부 입면 B / inner elevation B

1 WALL : APP. COLOR LACQ' FIN. / WALL : T18 MDF 9MM JOINT ON APP. COLOR LACQ' FIN. **2** TRACK LIGHTING **3** IMAGE WALL : APP. DUE DILIGENCE COMPUTER CUTTING INSTALLING / LIGHT BOX **4** WALL : T18 MDF ON APP. COLOR LACQ' FIN. **5** HANGER : R15 PIPE ON APP. COLOR LACQ' FIN. **6** INDIRECT LIGHTING FL32W / IMAGE WALL : APP. DUE DILIGENCE COMPUTER CUTTING INSTALLING **7** T9 MDF ON APP. COLOR LACQ' FIN. **8** T12 CLEAR GLASS **9** T65 STUD **10** T65 RUNNER **11** T9 MDF **12** 38X38 SQUARE WOOD

1 벽 : 지정 컬러 래커 마감 / 벽 : 지정 컬러 래커 마감 위 T18 MDF 9MM 매지 **2** 트랙 조명 **3** 이미지 벽 : 지정 상세 컴퓨터 커팅 취부 / 조명 박스 **4** 벽 : 지정 컬러 래커 마감 위 T18 MDF **5** 옷걸이 : 지정 컬러 래커 마감 위 R15 파이프 **6** 간접 조명 FL32W / 이미지 벽 : 지정 상세 컴퓨터 커팅 취부 **7** 지정 컬러 래커 마감 위 T9 MDF **8** T12 투명 유리 **9** T65 스터드 **10** T65 러너 **11** T9 MDF **12** 38X38 각재

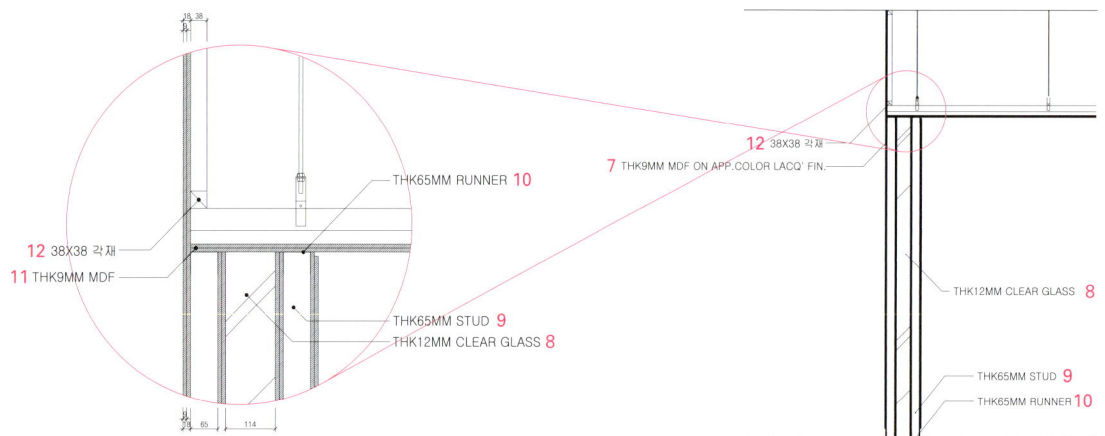

회의실 단면 G / conference room section G

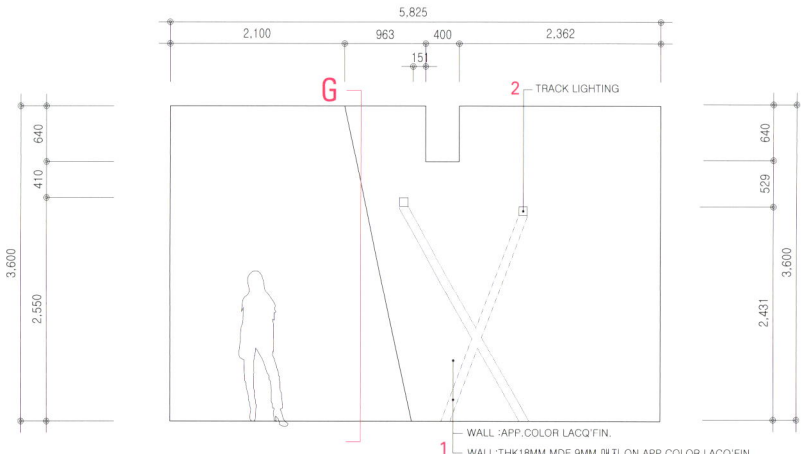

회의실 입면 D / conference room elevation D

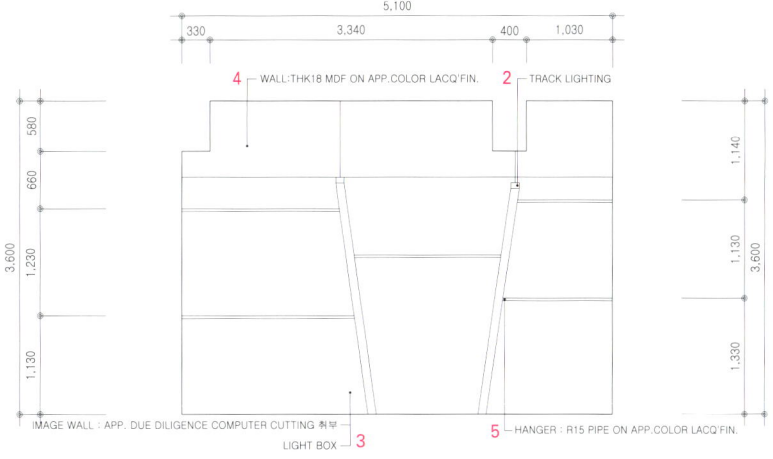

회의실 입면 E / conference room elevation E

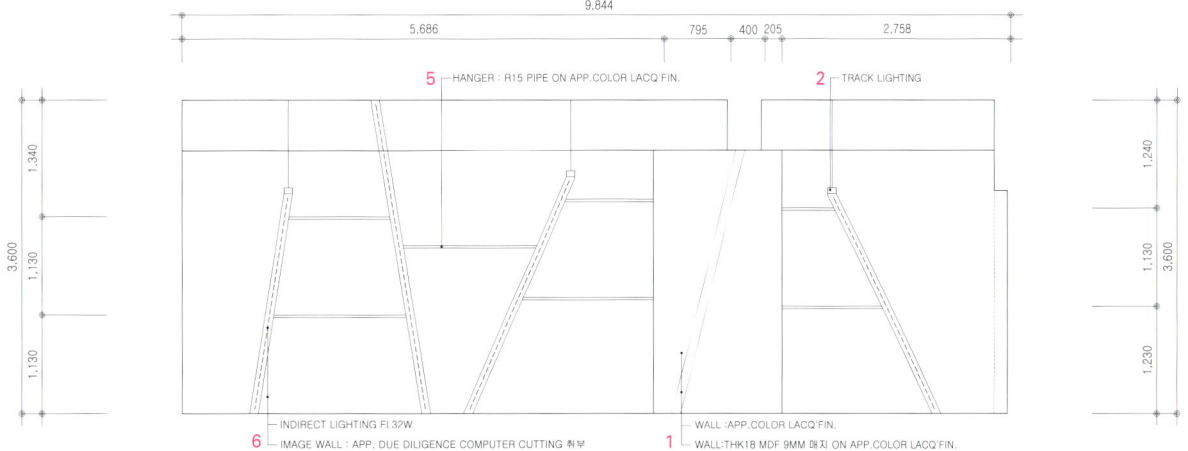

회의실 입면 F / conference room elevation F

JAGOK-DONG OFFICE

로담 에이아이 | 김영옥 Rodemn AI | Kim Young Ok

본 프로젝트는 로담 에이아이의 자곡동 사옥이다. 마치 백색의 상자가 쌓여있는 듯한 볼륨의 형태가 가까이의 청계산과 도시 콘텍스트와 어울려 색다른 분위기를 만들어 낸다. 하부의 큰 볼륨에서 위층으로 올라갈수록 작은 볼륨이 적층되는 형태이다. 내부는 단순한 박스 형태의 외관과는 다르게 풍부한 공간이 구성되었다. 외벽으로 감싼 듯 보이는 1층 필로티 공간은 내부이지만 외부 공간의 성격을 동시에 가진다. 2층의 도서관 볼륨은 2개 층을 통과하는 높은 천장고와 독특한 형태로 내부 공간을 풍부하게 한다. 특별한 재료의 질감이나, 독특한 색상 없이 장식 없는 단순한 형태지만 서로 소통하는 내부 공간, 외부와 내부의 긴밀한 구성이 프로젝트를 화려하게 한다.

The project is to construct the office building of Rodemn AI in Jagok-dong. The building is differentiated in the urban context near Mt. Cheonggye as it looks like white boxes are filed up. The big volume of the bottom tapers as it rises. And the interior is luxuriously decorated unlikely the simple box - shaped exterior. The pilotis space on the 1st floor, which seems to be embraced by outer walls is eclectic while the library volume on the 2nd floor enriches the interior with high ceiling penetrating two layers and its distinctive structure. Despite of no special material, texture, color or ornament, the intimate communication between the exterior and interior dazzles the simple structured project.

위치 서울특별시 강남구 자곡동 291-2
용도 업무
면적 186m²
규모 3층
설계기간 2008. 6 ~ 10
공사기간 2009. 2 ~ 2010. 8
설계팀 백현우, 서정용, 이유란, 최성우, 허진설
사진 김재윤

Location 291-2, Jagok-dong, Gangnam-gu, Seoul
Use Office
Area 186m²
Building scope 3F
Design period 2008. 6 ~ 10
Construction period 2009. 2 ~ 2010. 8
Photographer Kim Jae Youn

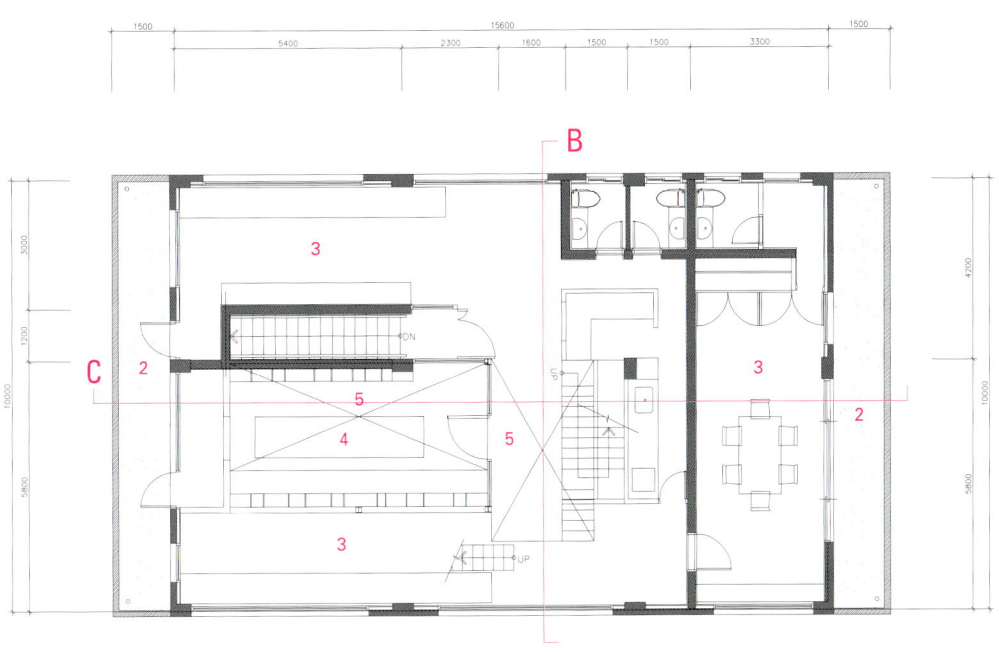

2층 평면도 / 2nd floor plan

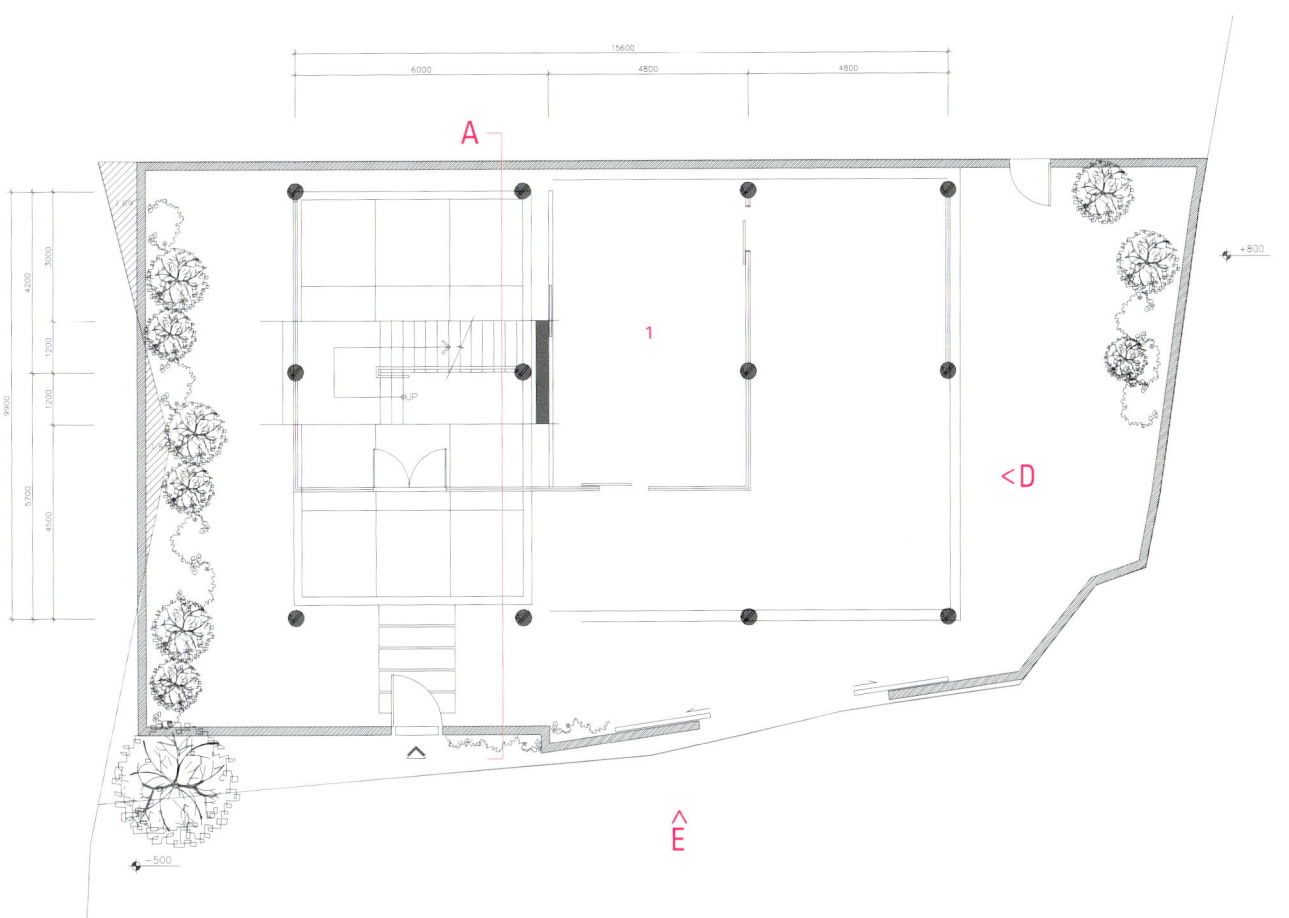

1층 평면도 / 1st floor plan

1 STORAGE 2 TERRACE 3 OFFICE 4 LIBRARY 5 ABOVE OPEN

1 창고 2 테라스 3 사무실 4 도서관 5 상부 열림

1 SHEET FIN. **2** CONCRETE W/ V.P FIN. / T30 BIRCH PLYWOOD **3** T30 BIRCH PLYWOOD **4** T9.5 GYPSUM BOARD 2PLY / T9 PLYWOOD **5** CONCRETE W/ V.P FIN. **6** CONCRETE W/ V.P FIN. / APP. SHEET FIN.

1 시트 마감 **2** 콘크리트 및 비닐 페인트 마감 / T30 자작나무 합판 **3** T30 자작나무 합판 **4** T9.5 석고보드 2겹 / T9 합판 **5** 콘크리트 및 비닐 페인트 마감 **6** 콘크리트 및 비닐 페인트 마감 / 지정 시트 마감

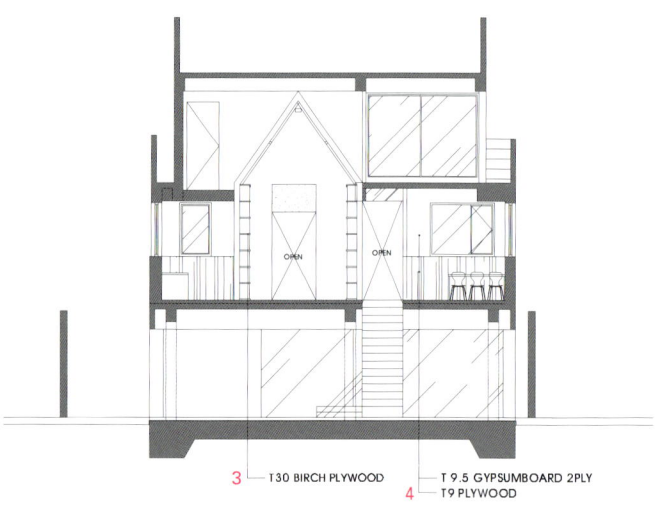

단면 A / Section A

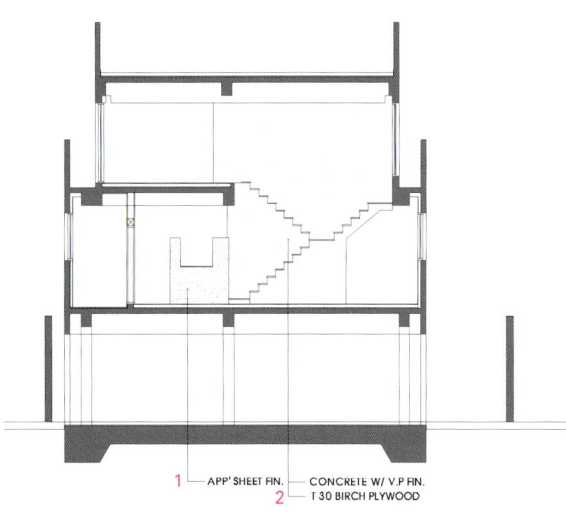

단면 B / Section B

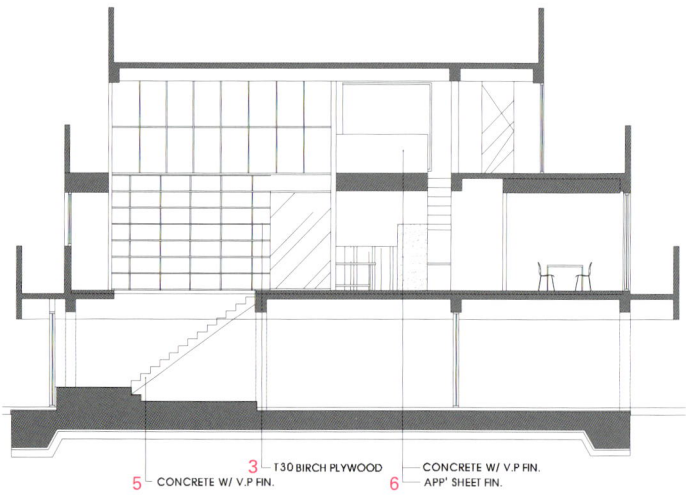

단면 C / Section C

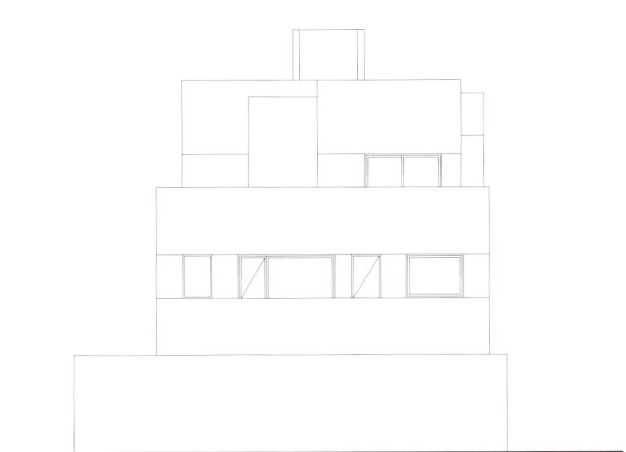

우측면도 D / right elevation D

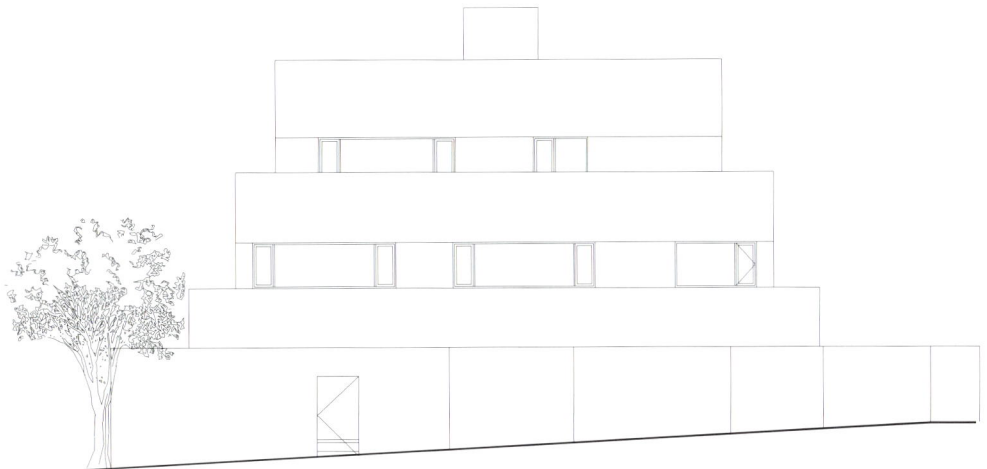

정면도 E / front elevation E

DMC CJ ENTERTAINMENT & MEDIA CENTER

(주)비안디자인 | 안경두 Beyond Design | Ahn Kyung Doo

DMC CJ Entertainment & Media Center는 아시아 최대 엔터테인먼트 회사인 CJ그룹의 영화 제작, 인터넷 게임, CGV, Mnet을 비롯한 수십 개의 케이블 TV 채널 제작 계열사들로 이루어져 있다. 저층부는 미디어 전시관, 오픈 스튜디오, 멀티 스튜디오, 미디어 홍보관, 다목적홀, 신인 인큐베이팅실 등 다양한 전시, 홍보, 체험, 상영, 공연, 촬영 시설들로 구성되어 있다. 주요한 공간 개념은 'Backstage'로 편집된 컨텐츠가 아닌, 생성되고 있는 과정 그대로의 컨텐츠가 노출되는 공간을 의미한다. 공용 공간과 동선 공간들은 컨텐츠 생성 공간과 연계되고, 내부 스튜디오나 공연 공간에서 생성되는 컨텐츠들은 공용 부위의 공간을 구성하는 요소로 반영된다.

DMC CJ Entertainment & Media Center is a headquarter of the largest entertainment company in Asia including CJ Entertainment, CJ Internet, CGV, Mnet, etc. The lower sector of the building consists of several creative facilities such as Media Exhibition Hall, Multi Studio, Open Studio, Multi - purpose Auditorium, Incubating UCC Room, and Restroom Gallery. Main concept of the project starts from 'Backstage'. The place should expose the contents in process without planned editing. Visitors will see and feel the power of the vivid raw contents.

위치 서울특별시 마포구 상암동 디지털미디어시티
용도 복합
마감 바닥 – 우드 플로링, 천연 데크재, 대리석, 타일, 대리석 타일 / 벽 – 흑경, 블랙 실버 컬러 스텐, 노출콘크리트, 화강석, 목재 타공 흡음판(라인형), LED 벽 / 천장 – 우드 루버, 노출 격자 루버(블랙도장, 우드 사출), 바리솔, 브론즈 컬러 스텐, 비닐 페인트, LED 천장
설계기간 2008. 7 ~ 2009. 12
공사기간 2009. 4 ~ 2010. 3
건축주 CJ
디자인팀 서윤희, 김주영, 최정훈, 이혜진, 차윤진, 황수익
사진 (주)비안디자인 제공

Location Digital Media City, Sangam-dong, Mapo-gu, Seoul
Use Complex
Finishing Floor - Wood flooring, Wood deck, Marble, Tile, Marble tile / Wall - Black mirror, Black silver s'stl plate, Exposed concrete, Granite, Accustic wood panel, LED board wall / Ceiling - Wood louver, Exposed grid louver(Black lac'q paint, Extruded wood), Barrisol, Brass s'stl plate, Vinyl paint, LED board ceiling
Design period 2008. 7 ~ 2009. 12
Construction period 2009. 4 ~ 2010. 3
Photos offer Beyond Design

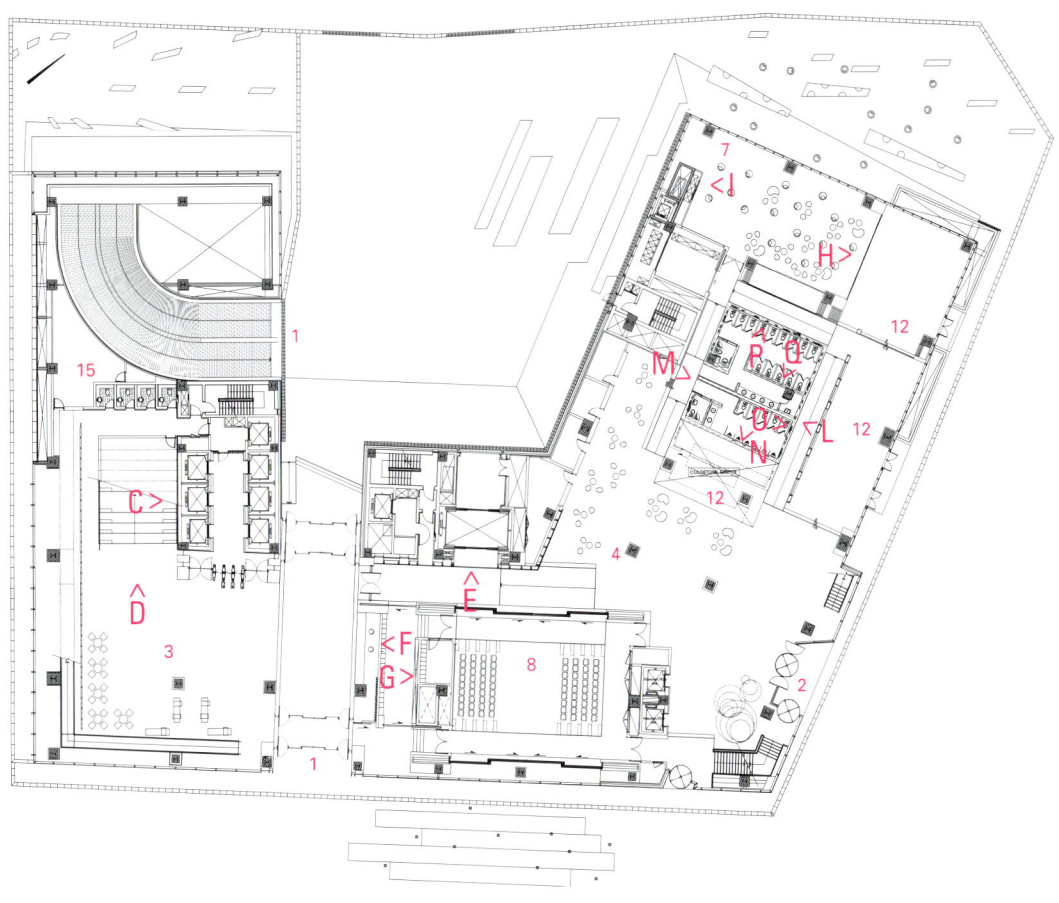

1층 평면도 / 1st floor plan

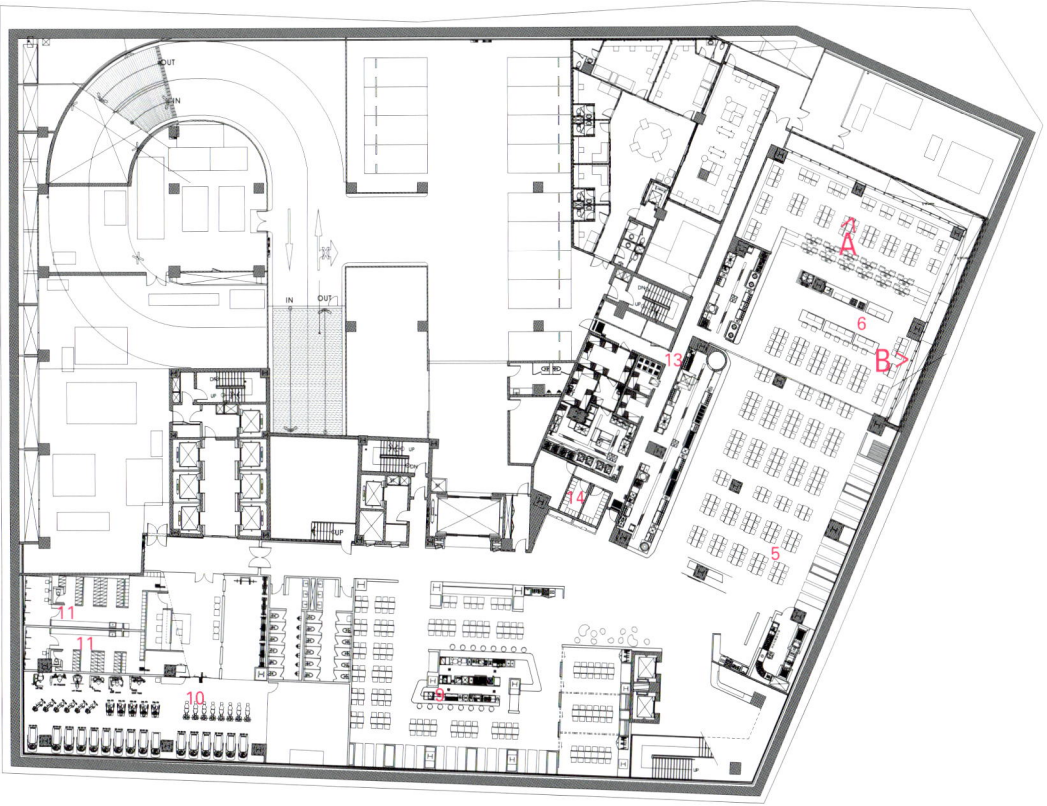

지하 1층 평면도 / B1F floor plan

1 MAIN ENTRANCE 2 SUB ENTRANCE 3 LOBBY 4 OPEN LOUNGE 5 HALL 6 OPEN STUDIO 7 SWING STUDIO 8 BACKSTAGE 9 OPEN BAR 10 FITNESS CENTER 11 LOCKER ROOM 12 RETAIL 13 WAITING ROOM / KITCHEN 14 KITCHEN / CLEAN ROOM 15 OFFICE

1 주출입구 2 부출입구 3 로비 4 오픈 라운지 5 홀 6 오픈 스튜디오 7 스윙 스튜디오 8 백스테이지 9 오픈 바 10 피트니스 센터 11 로커룸 12 판매점 13 출연자 대기실 / 주방 14 주방 / 세척실 15 사무실

1 MAIN ENTRANCE 2 LOBBY 3 MEDIA GALLERY 4 LOUVER : APP. VENEER WOOD FIN. 5 BANQUET SEAT 6 CON'C FIN. 7 APP. DARK BROWN FABRIC FIN. 8 INNER PART OF LOUVER APP. BLACK PAINTING FIN. 9 APP. TABLE / APP. BLACK COLOR STS FIN. 10 APP. VENEER WOOD PANEL FIN. / APP. BLACK PAINTING FIN. ON H - TYPE METAL 11 CON'C PILLAR : APP. BLACK PAINTING 12 TRACK : APP. VENEER WOOD FIN. 13 METAL PIN : APP. BLACK PAINTING FIN. 14 APP. DARK BROWN FABRIC FIN. / APP. VENEER WOOD PANEL FIN.

1 주출입구 2 로비 3 미디어 갤러리 4 루버 : 지정 무늬목 패널 마감 5 연회석 6 콘크리트 마감 7 지정 다크 브라운 패브릭 마감 8 루버 안쪽 지정 블랙 도장 마감 9 지정 테이블 / 지정 블랙 컬러 스텐 마감 10 지정 무늬목 패널 마감 / H형 금속 위 지정 블랙 도장 마감 11 콘크리트 기둥 : 지정 블랙 도장 12 트랙 : 지정 무늬목 마감 13 금속 핀 : 지정 블랙 도장 마감 14 지정 다크 브라운 패브릭 마감 / 지정 무늬목 패널 마감

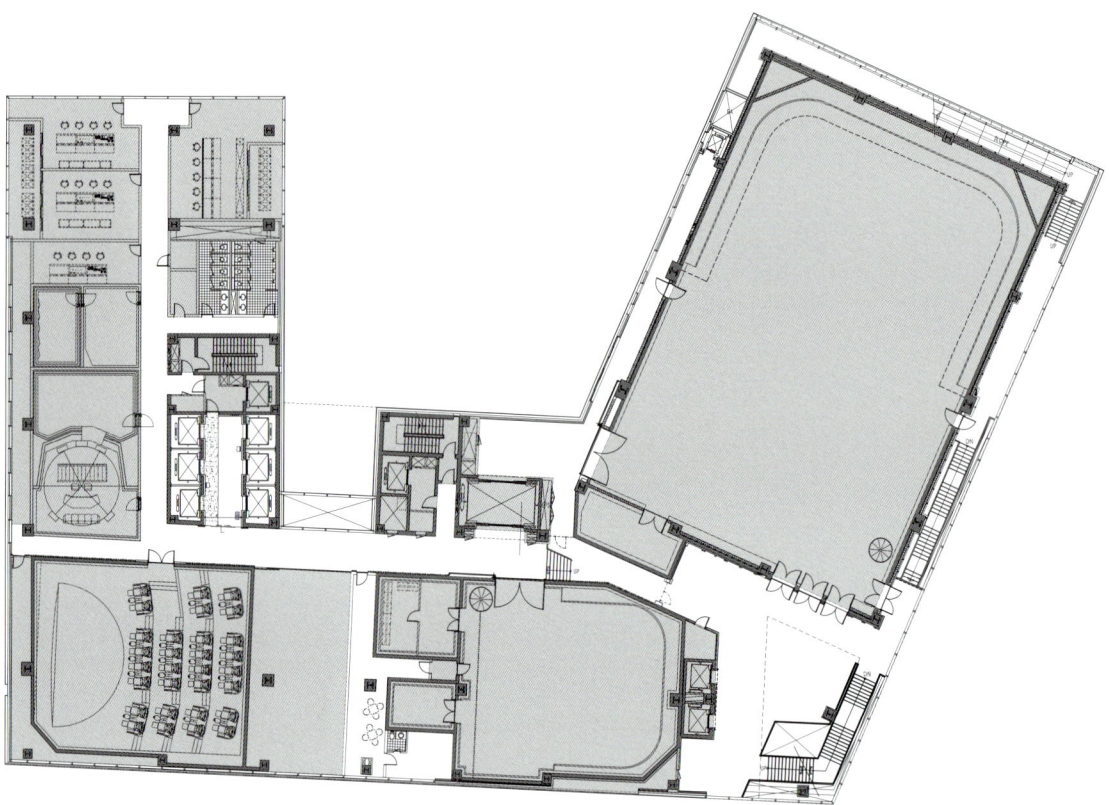

2층 평면도 / 2nd floor plan

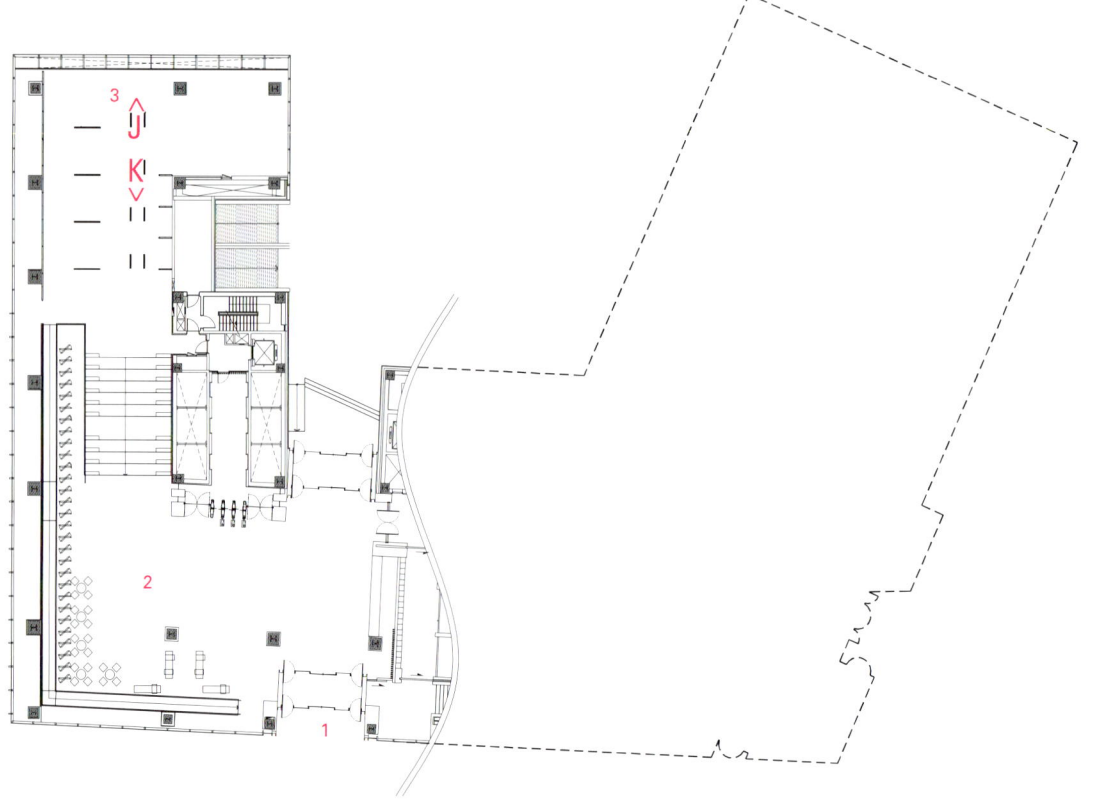

1.5층 평면도 / 1.5F floor plan

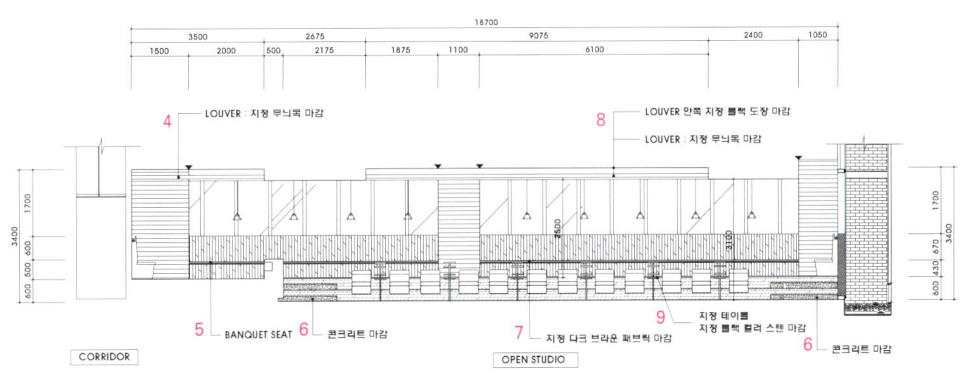

푸드 코트 입면 A / food court elevation A

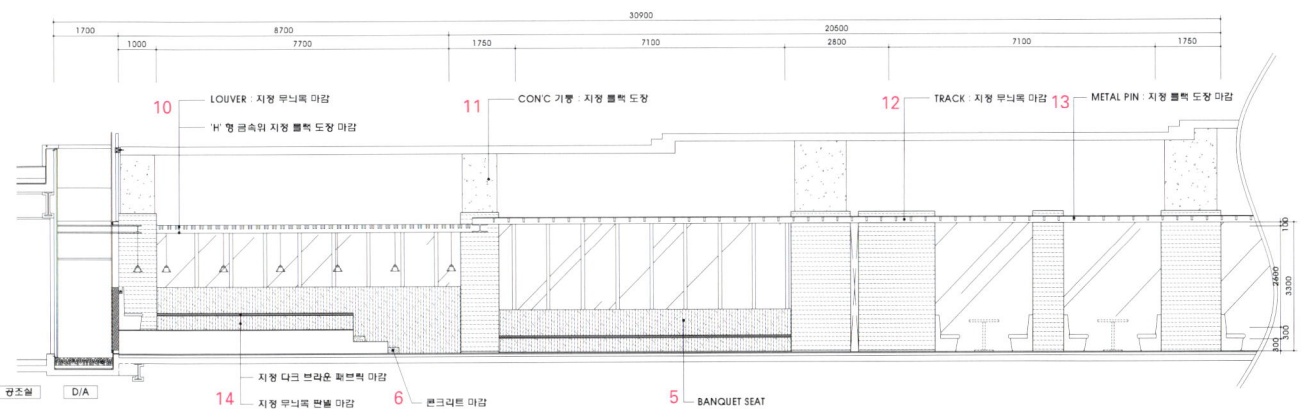

푸드 코트 입면 B / food court elevation B

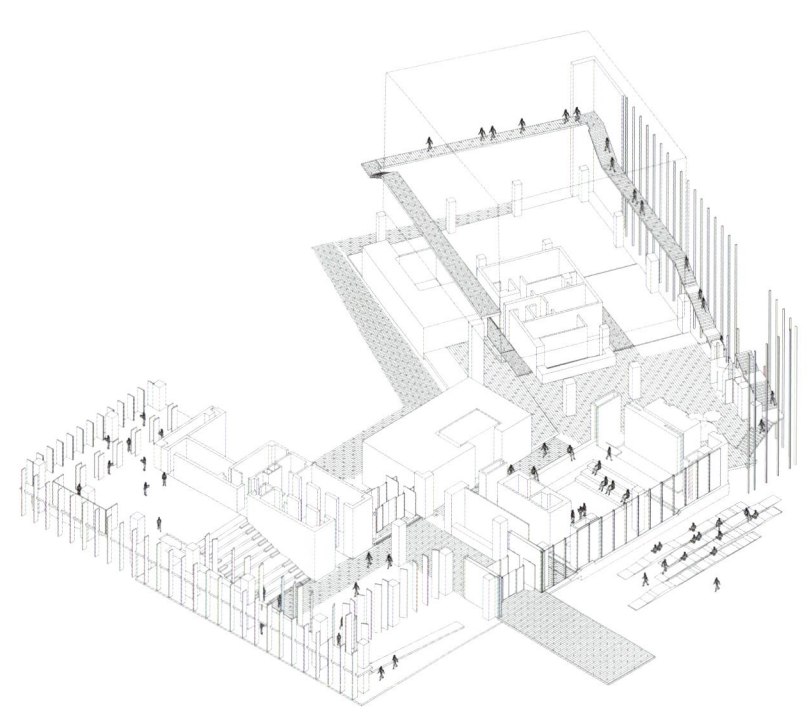

1층 입체도 / 1st floor axonometric

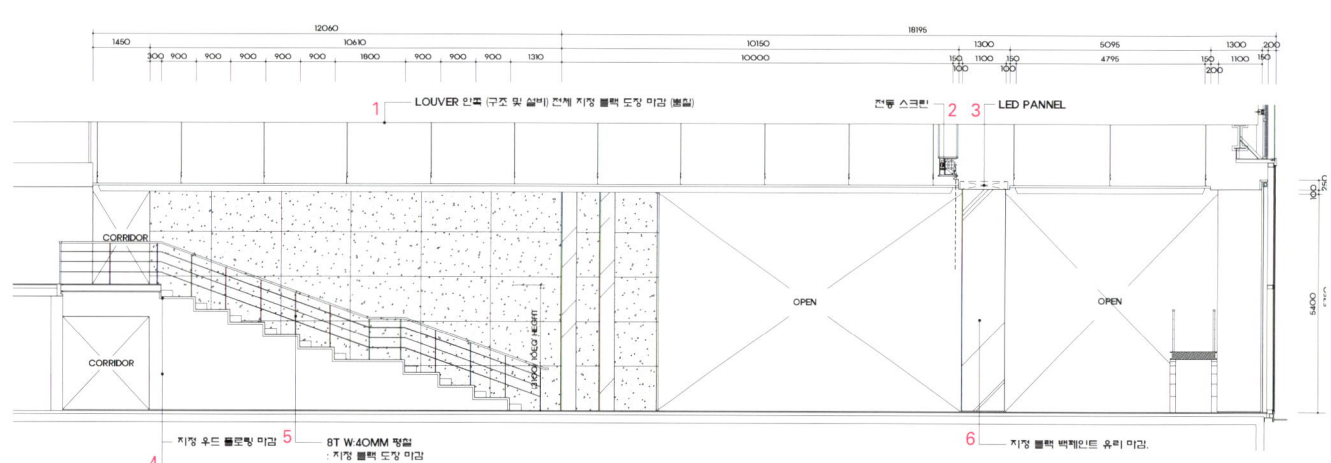

로비 입면 C / lobby elevation C

1 INNER PART OF LOUVER : APP. BLACK PAINTING FIN.(SPRAYING) 2 ELECTRIC SCREEN 3 LED PANEL 4 APP. WOOD FLOORING FIN. / APP. WHITE PAINTING FIN. 5 8T W:40 FLAT STEEL : APP. BLACK PAINTING FIN. 6 APP. BLACK BACK PAINT GLASS FIN. 7 APP. BLACK GLOSSY BARRISOL FIN. 8 POST(SLOP STRUCTURE SUPPORT) / APP. BLACK PAINTING FIN. 9 EVOLVING WALL : FRONT SIDE - ACUSTIC WOOD PANEL / BACK SIDE - DETACHABLE METAL PANEL : APP. WHITE PAINTING / APP. STOPPER INSTALLING 10 APP. STONE FIN.

1 루버 안쪽 전체 지정 블랙 도장 마감(뿜칠) 2 전동 스크린 3 LED 패널 4 지정 우드 플로링 마감 / 지정 화이트 도장 마감 5 8T W:40 평철 : 지정 블랙 도장 마감 6 지정 블랙 백페인트 유리 마감 7 지정 블랙 유광 바리솔 마감 8 포스트(슬롭 구조 지지) / 지정 블랙 도장 마감 9 회전 벽 : 정면 – 어쿠스틱 우드 패널 / 후면 – 분리 가능한 금속 패널 : 지정 화이트 도장 / 지정 스토퍼 취부 10 지정 석재 마감

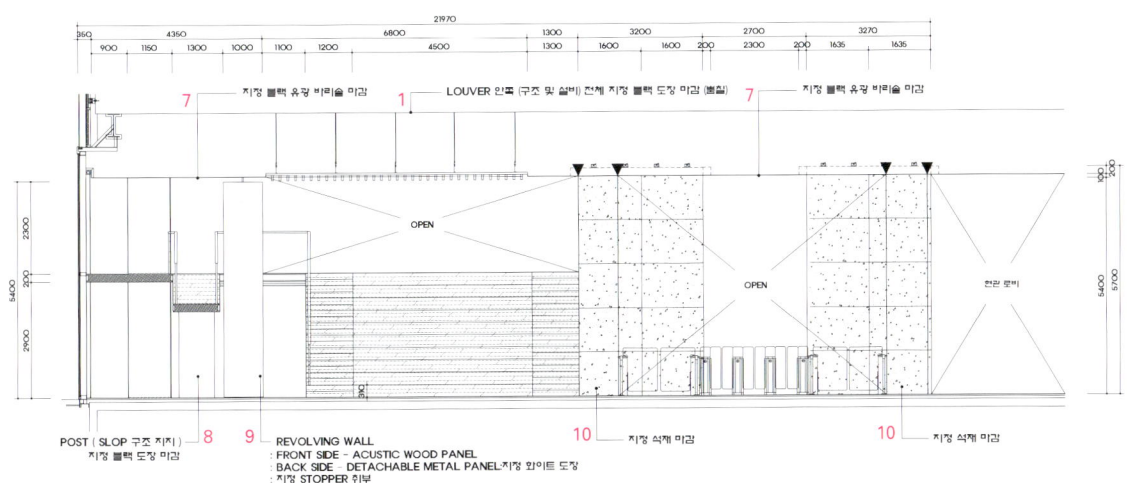

로비 입면 D / lobby elevation D

DMC CJ ENTERTAINMENT & MEDIA CENTER | DMC CJ 엔터테인먼트 & 미디어 센터

1 APP. LCD PANEL / BACK SIDE WALL : APP. BLACK BACK PAINT GLASS FIN. 2 CEILING WOOD TRACK EXTENSION 3 DOOR : APP. BLACK PAINTING FIN. 4 APP. BLACK PAINTING FIN. / CEILING WOOD TRACK EXTENSION 5 APP. LCD PANEL 6 INNER PART OF LOUVER : APP. BLACK PAINTING FIN.(SPRAYING) 7 APP. BLACK PAINTING FIN. 8 ACOUSTIC WOOD PANEL 9 MEMORY BOX 10 APP. BLACK COLOR STS FIN. 11 APP. VENEER WOOD FIN. 12 18MM WOOD LATTICE FRAME 13 MAIL BOX(APP. BLACK COLOR STS FIN.) 50MM SETBACK

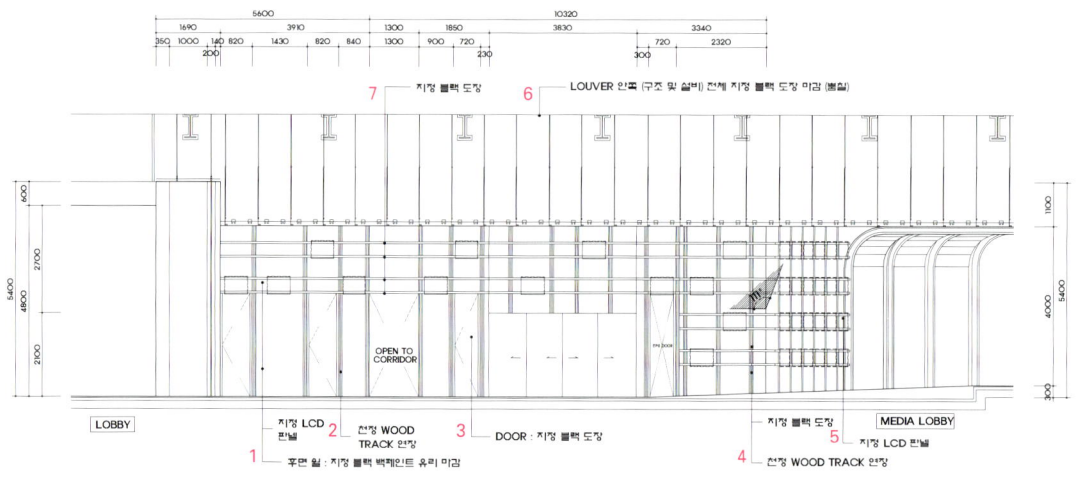

백스테이지 입면 E / backstage elevation E

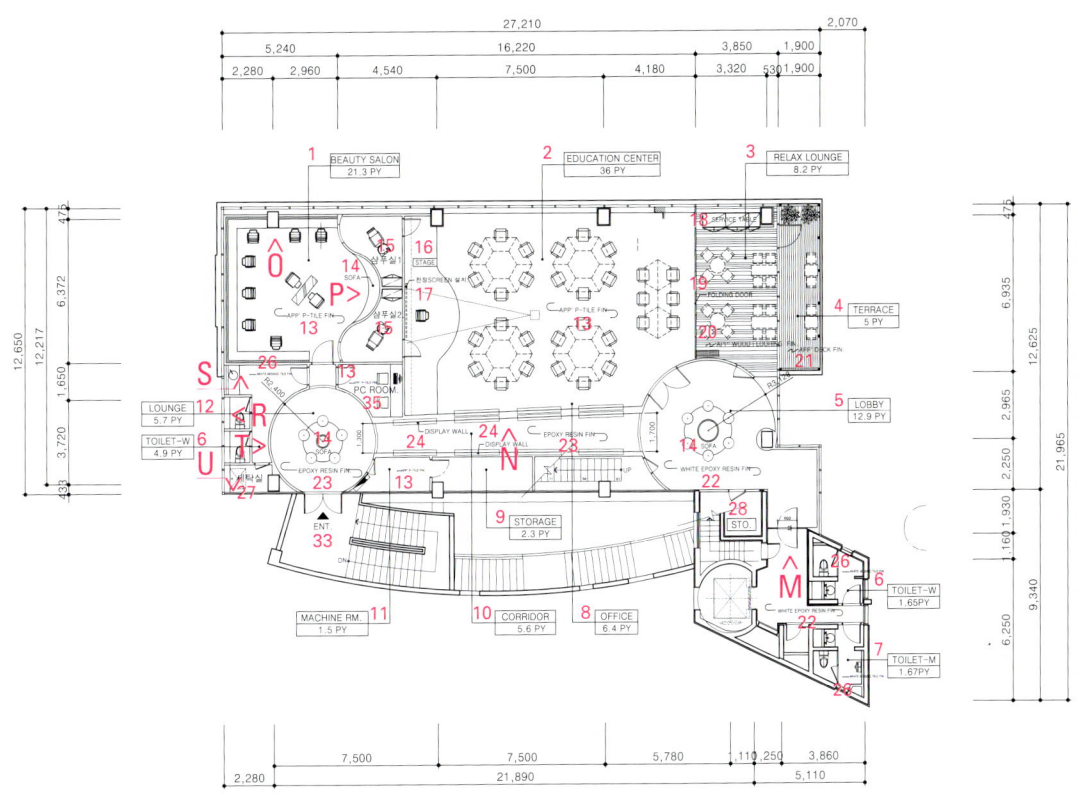

4층 평면도 / 4th floor plan

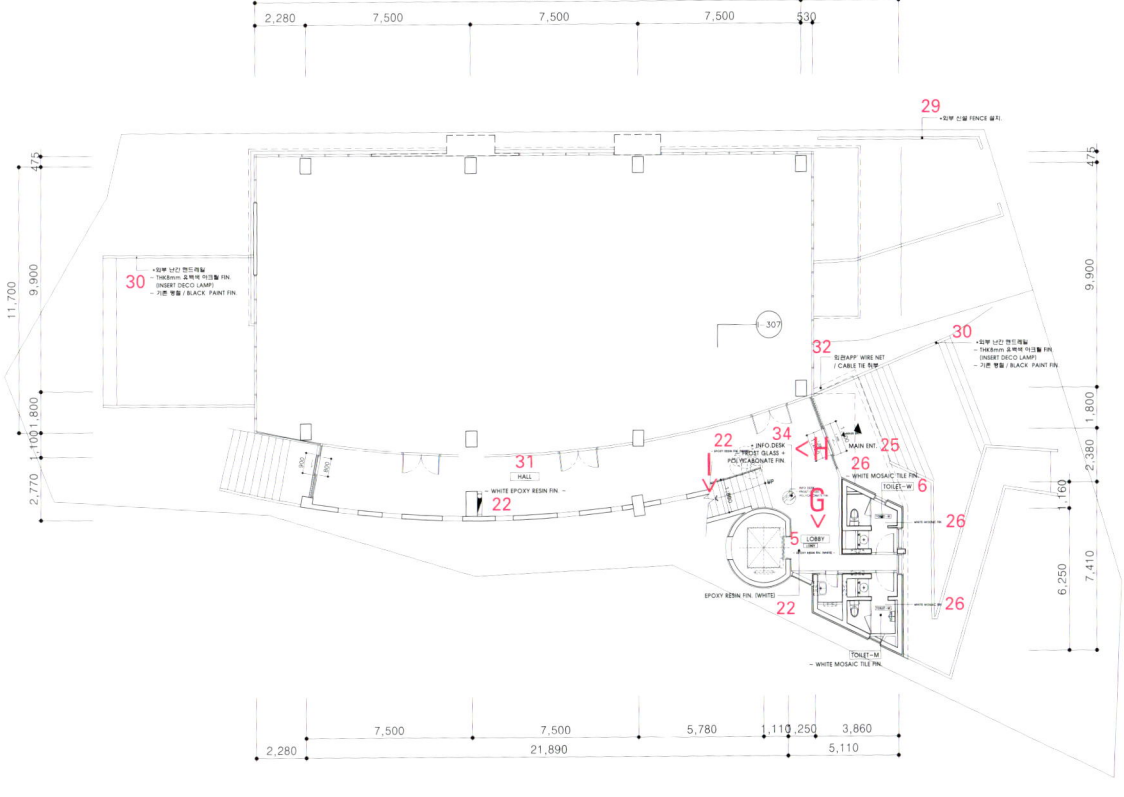

1층 평면도 / 1st floor plan

1 BEAUTY SALON 2 EDUCATION CENTER 3 RELAX LOUNGE 4 TERRACE 5 LOBBY 6 TOILET - W 7 TOILET - M 8 OFFICE 9 STORAGE 10 CORRIDOR 11 MACHINE RM. 12 LOUNGE 13 APP. P - TILE FIN. 14 SOFA 15 SHAMPOO RM 16 STAGE 17 CEILING SCREEN INSTALLATION 18 SERVICE TABLE 19 FOLDING DOOR 20 APP. WOOD FLOORING 21 APP. DECK FIN. 22 WHITE EPOXY RESIN FIN. 23 EPOXY RESIN FIN. 24 DISPLAY WALL 25 MAIN ENT. 26 WHITE MOSAIC TILE FIN. 27 LAUNDRY 28 STO. 29 OUTSIDE NEW FENCE INSTALLATION 30 OUTSIDE HANDRAIL : /T8 WHITE ACRYL FIN.(INSERT DECO LAMP), PLAT IRON, BLACK PAINT FIN. 31 HALL 32 EXTERIOR APP. WIRE NET, CABLE TILE INSTALLATION 33 ENT. 34 INFO. DESK FROST GLASS, POLYCABONATE FIN. 35 PC ROOM

1 뷰티샵 2 교육 센터 3 휴게 라운지 4 테라스 5 로비 6 화장실(여) 7 화장실(남) 8 사무실 9 창고 10 복도 11 기계실 12 라운지 13 지정 P – 타일 마감 14 소파 15 샴푸실 16 무대 17 천정 스크린 설치 18 서비스 테이블 19 접이문 20 지정 우드 플로링 21 지정 데크 마감 22 흰색 에폭시 수지 마감 23 에폭시 수지 마감 24 디스플레이 벽 25 메인 출입구 26 흰색 모자이크 타일 마감 27 세탁실 28 창고 29 외부 신설 울타리 설치 30 외부 난간 핸드레일 : /T8 유백색 아크릴 마감(데코램프 삽입), 기전 평철, 블랙 페인트 마감 31 홀 32 외관 지정 철망, 케이블 타일 취부 33 입구 34 안내 데스크/서리 유리, 폴리카보네이트 마감 35 PC실

1 APP. WIRE NET, WHITE CABLE TILE INSTALLATION(INSERT LIGHTING - INSIDE LIGHT PROJECTOR INSTALLATION) 2 SIGN - NEON INSTALLATION 3 SIGN - NEON INSTALLATION / CANOPY MAKING : T1.6 GALVA, BLACK METAL PAINT. 4 SIGN LOGO SCASI, INSIDE - NEON LIGHTING 5 CANOPY : BLACK PAINT FIN. 6 HALOGEN DOWN LIGHTING(OUTSIDE TYPE) 7 MAIN ENT. 8 SIGN : T5 ACRYLIC FIN.(INSERT LIGHTING) 9 AUTO DOOR : BLACK PAINT FIN. 10 SQUARED PIPE REINFORCEMENT, APP. WIRE NET, WHITE CABLE TILE INSTALLATION(INSERT LIGHTING - INSIDE LIGHT PROJECTOR INSTALLATION) 11 T9 STEEL PLATE MAKING SLOPE / BLACK BAKING PAINTING(T1.6 GALV, BLACK PAINT FIN.) 12 LOWER CANOPY MAKING : SQUARED PIPE REINFORCEMENT, T1.6 GALVA, BLACK METAL PAINT 13 HANDRAIL : T8 WHITE ACRYL(INSERT DECO LAMP.) / EXISTING FLAT STEEL, BLACK PAINT FIN. 14 SQUARED PIPE REINFORCEMENT STRUCTURE FRAME INSTALLATION / INSIDE SIGN - NEON INSTALLATION / APP. WIRE NET, WHITE CABLE TILE INSTALLATION(INSERT LIGHTING - INSIDE LIGHT PROJECTOR INSTALLATION) 15 OUTSIDE SIGN - NEON INSTALLATION 16 HALOGEN DOWN LIGHTING(OUTSIDE TYPE)

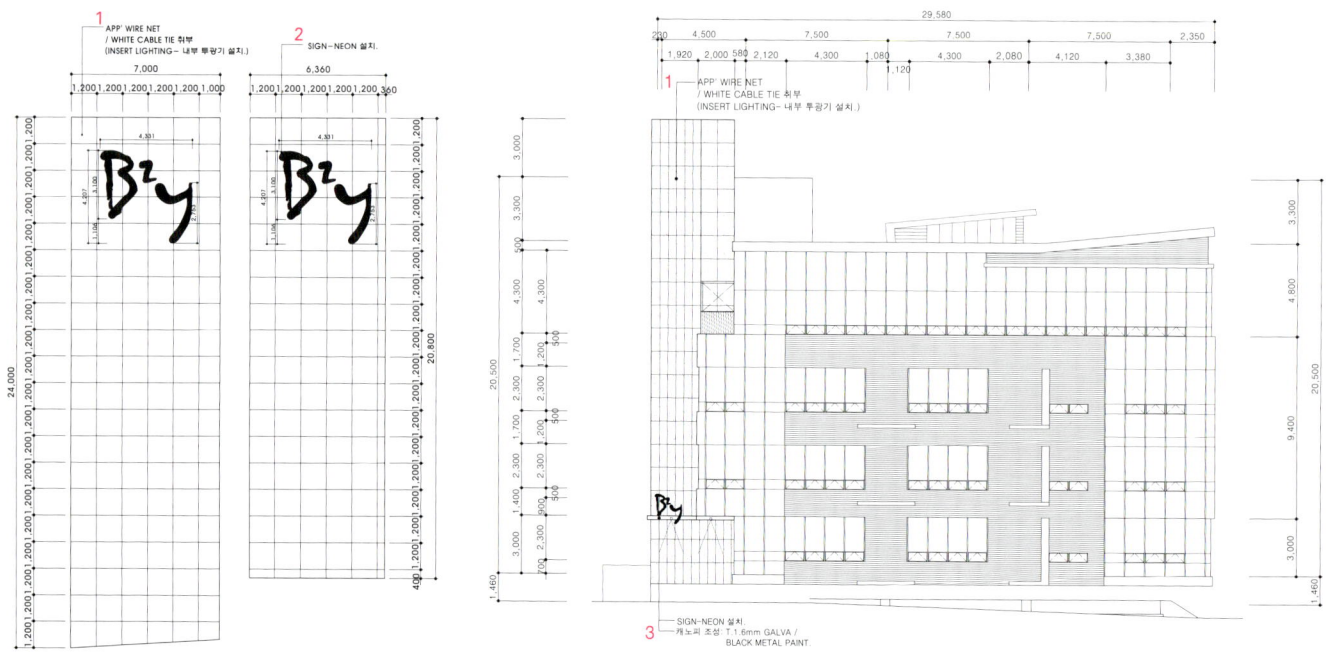

파사드 사인 상세 / facade sign detail

파사드 입면 / facade elevation

1 지정 철망, 흰색 케이블 타일 취부(삽입 조명 – 내부 투광기 설치) 2 사인 – 네온 설치 3 사인 – 네온 설치 / 캐노피 조성 : T1.6 갈바, 블랙 메탈 페인트 4 사인 로고 스카시, 내부 – 네온 조명 5 캐노피 : 블랙 페인트 마감 6 할로겐 다운라이트(외부용) 7 주 출입구 8 사인 : T5 아크릴 마감(삽입 조명) 9 자동문 : 블랙 페인트 마감 10 각파이프 보강, 지정 철망, 흰색 케이블 타일 취부(삽입 조명 – 내부 투광기 설치) 11 T9 철판 물구배주기 / 블랙 소부도장(T1.6 갈바, 블랙 페인트 마감) 12 하부 캐노피 조성 : 각파이프 보강, T1.6 갈바, 블랙 메탈 페인트 13 난간 : T8 유백아크릴(데코램프 삽입) / 기존 평철, 블랙 페인트 마감 14 각파이프 보강 구조틀 설치 / 내부 사인 – 네온 설치 / 지정 철망, 흰색 케이블 타일 취부(삽입 조명 – 내부 투광기 설치) 15 외부 사인 – 네온 설치 16 할로겐 다운라이트(외부용)

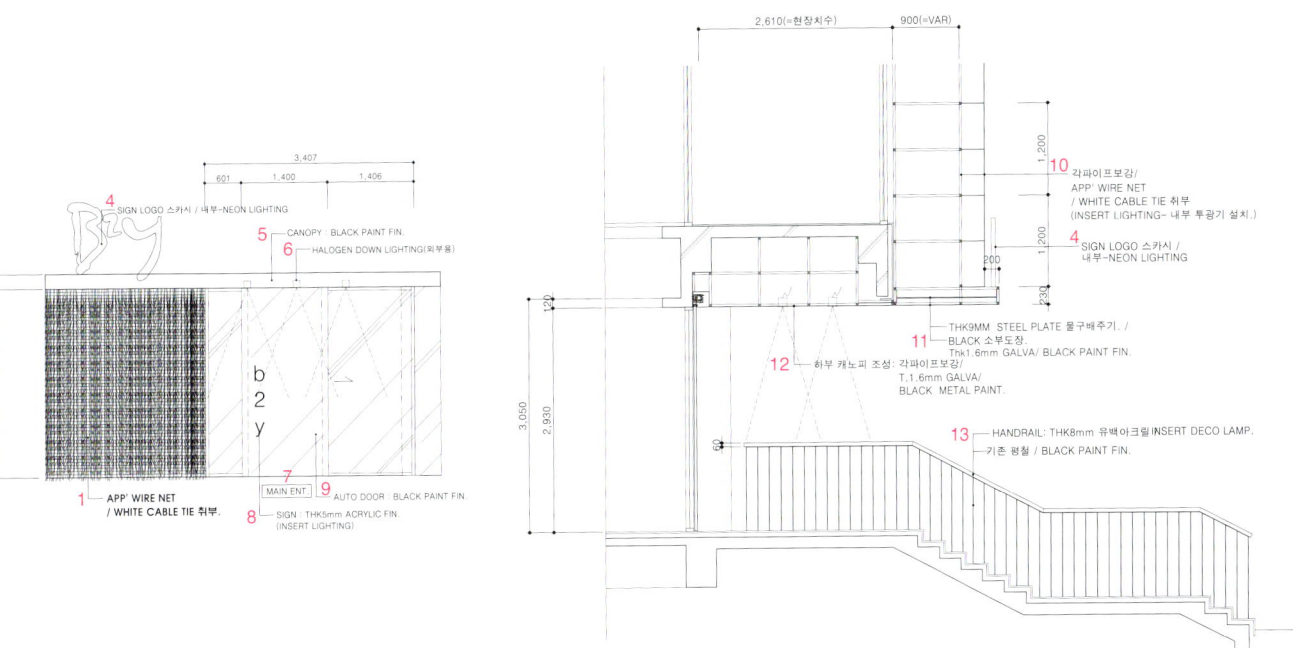

주출입구 입면 A / main entrance elevation A

주출입구 단면 B / main entrance section B

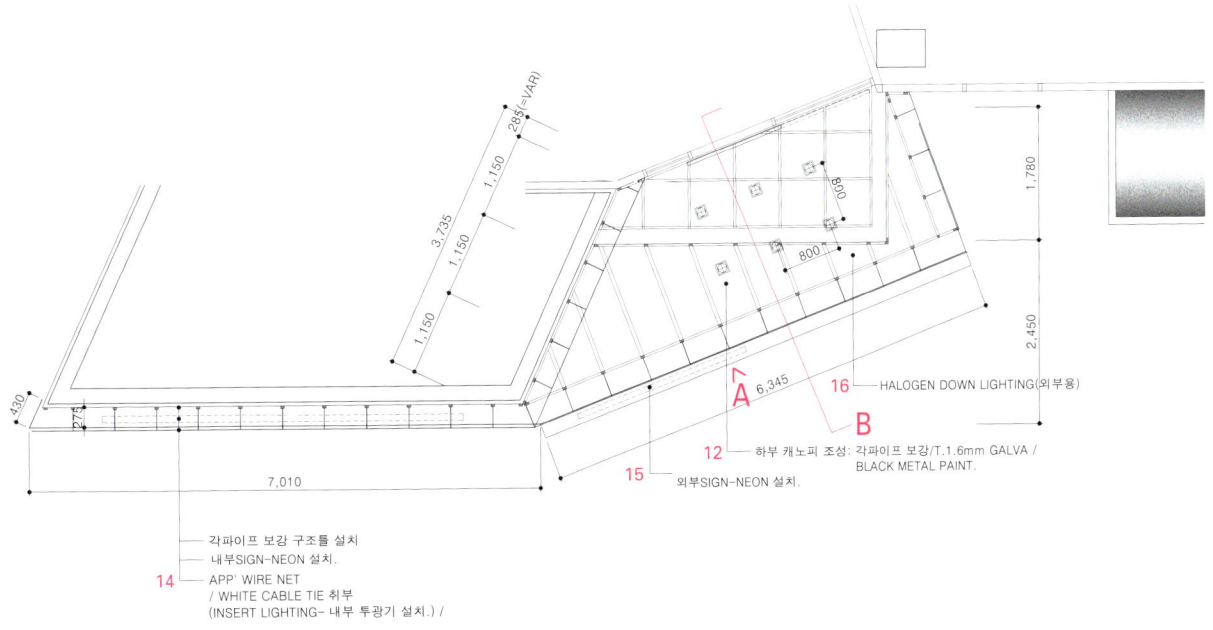

주출입구 부분 평면 / main entrance partial floor plan

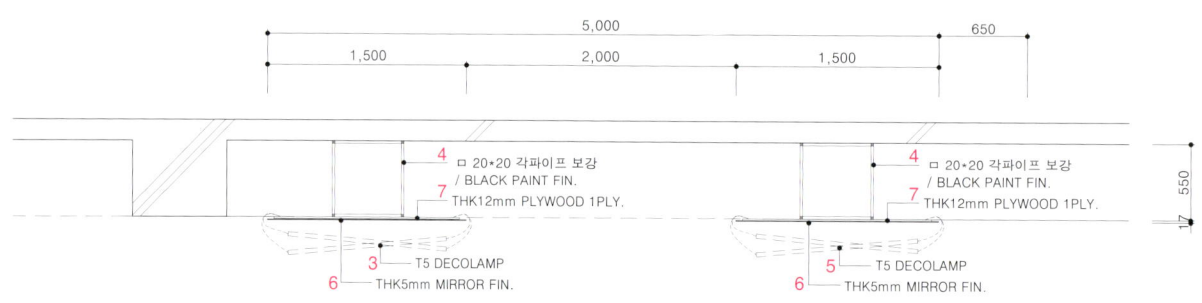

천장 상세 C / ceiling detail C

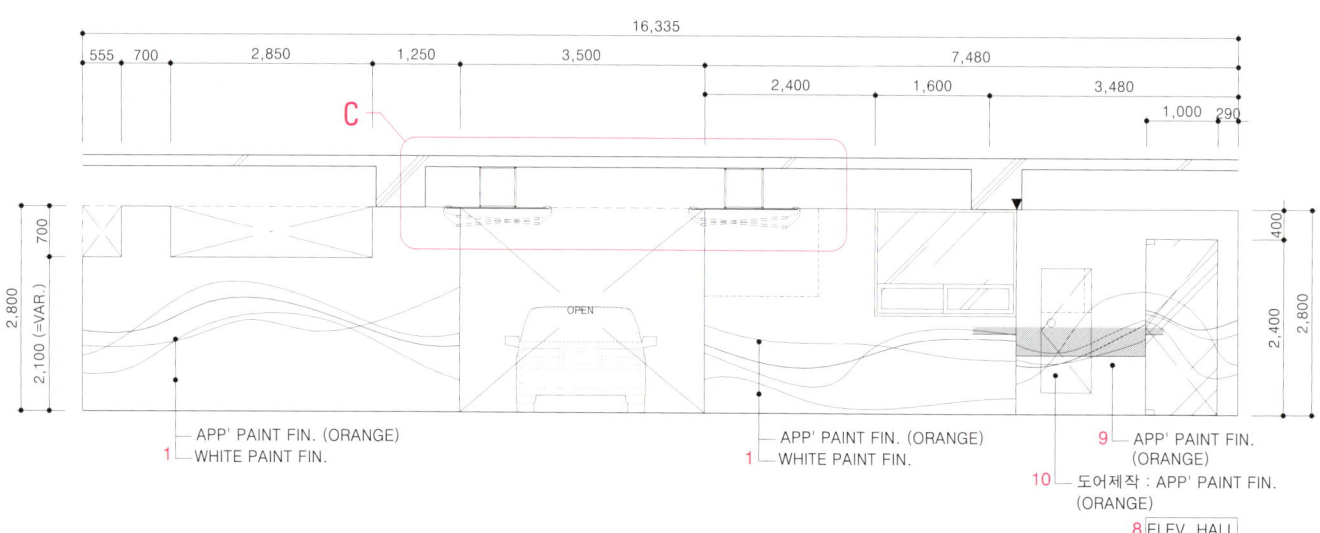

지하 1층 주차장 입면 / B1 floor parking lot elevation

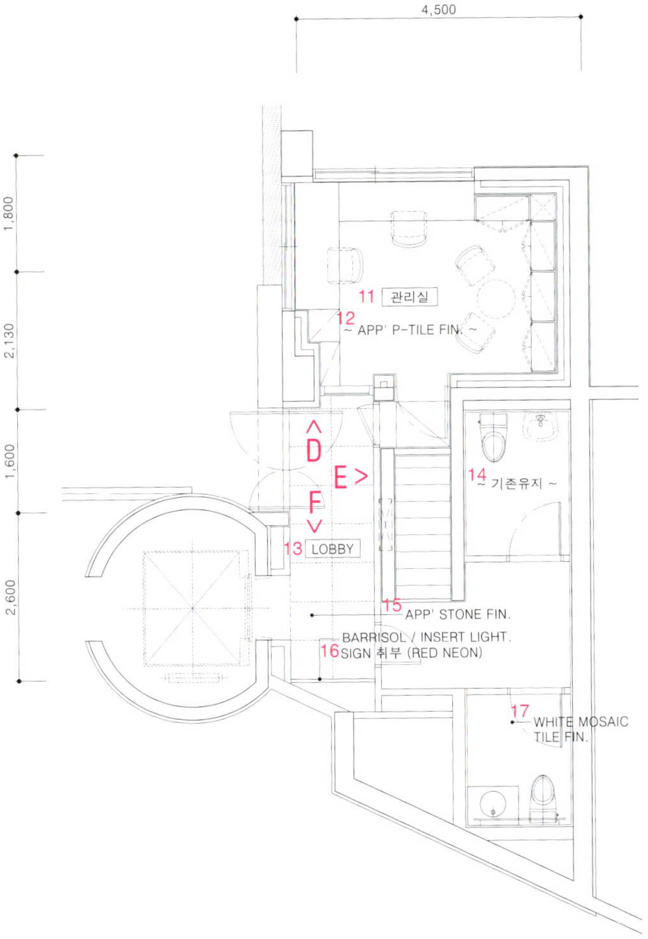

지하 1층 부분 평면 / B1 partial floor plan

1 APP. PAINT FIN.(ORANGE) / WHITE PAINT FIN. 2 APP. RED NEON / APP. BARRISOL FIN., INSERT LIGHTING 3 T5 DECOLAMP 4 □-20X20 SQUARED PIPE REINFORCEMENT, BLACK PAINT FIN. 5 T5 DECOLAMP 6 T5 MIRROR FIN. 7 T12 PLYWOOD 1PLY 8 ELEV. HALL 9 APP. PAINT FIN.(ORANGE) 10 MAKING DOOR : APP. PAINT FIN.(ORANGE) 11 SAFETY DIRECTORATE 12 APP. P - TILE FIN. 13 LOBBY 14 EXISTING MAINTENANCE 15 APP. STONE FIN. 16 BARRISOL, INSERT LIGHT SIGN INSTALLATION(RED NEON) 17 WHITE MOSAIC TILE FIN. 18 BLACK PAINT FIN. 19 TOILET / STO. 20 WHITE PAINT FIN. 21 GROOVE : BLACK PAINT FIN. 22 EXISTING CABINET PANEL

1 지정 페인트 마감(주황색) / 흰색 페인트 마감 2 지정 레드 네온 / 지정 바리솔 마감, 삽입 조명 3 T5 데코램프 4 □-20X20 각파이프 보강, 블랙 페인트 마감 5 T5 데코램프 6 T5 거울 마감 7 T12 합판 1겹 8 엘리베이터 홀 9 지정 페인트 마감(주황색) 10 도어제작 : 지정 페인트 마감(주황색) 11 관리실 12 지정 P-타일 마감 13 로비 14 기존 유지 15 지정 석재 마감 16 바리솔, 삽입 사인조명 취부(레드 네온) 17 흰색 모자이크 타일 마감 18 블랙 페인트 마감 19 화장실 / 창고 20 흰색 페인트 마감 21 홈줄 : 블랙 페인트 마감 22 기존 분전함

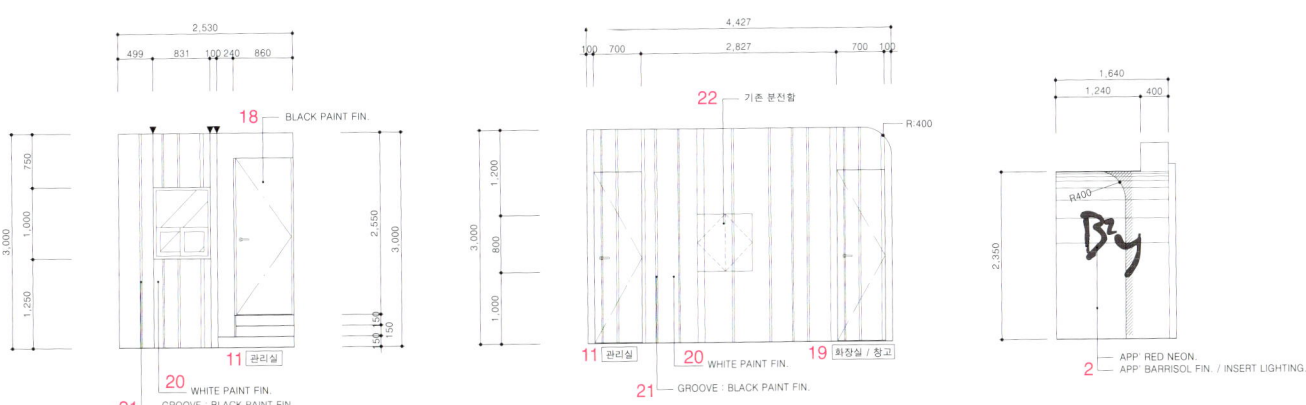

지하 1층 입면 D / B1 elevation D

지하 1층 입면 E / B1 elevation E

지하 1층 입면 F / B1 elevation F

1 APP. RING SHAPE FLUORESCENT LAMP(Ø=380) 2 UPPER PLATE : Ø620 T8 TEMPEREAD FROST GLASS 3 SIGN : APP. NEON 4 Ø600 T5 CLEAR ACRYL / FROST SHEET FIN. 5 T9 MDF, WHITE PAINT FIN. 6 T18 MDF, WHITE PAINT FIN. 7 MOSAIC DIRECTION 8 TOILET - M 9 STO. 10 GROOVE : T10, BLACK PAINT FIN. / BASE : H=30MM, BLACK PAINT FIN. 11 APP. PAINT FIN., BEADS CURTAIN WALL 12 OPEN 13 ELEV. 14 1F HALL

1 지정 환형형광등(Ø=380) 2 상판 : Ø620 T8 강화 서리 유리 3 사인 : 지정 네온 4 Ø600 T5 투명 아크릴 / 서리 시트 마감 5 T9 MDF, 흰색 페인트 마감 6 T18 MDF, 흰색 페인트 마감 7 알판와리 방향 8 화장실(남) 9 창고 10 홈줄 : T10, 블랙 페인트 마감 / 베이스 : H=30MM, 블랙 페인트 마감 11 지정 페인트 마감, 비즈 커튼월 12 열림 13 엘리베이터 14 1F 홀

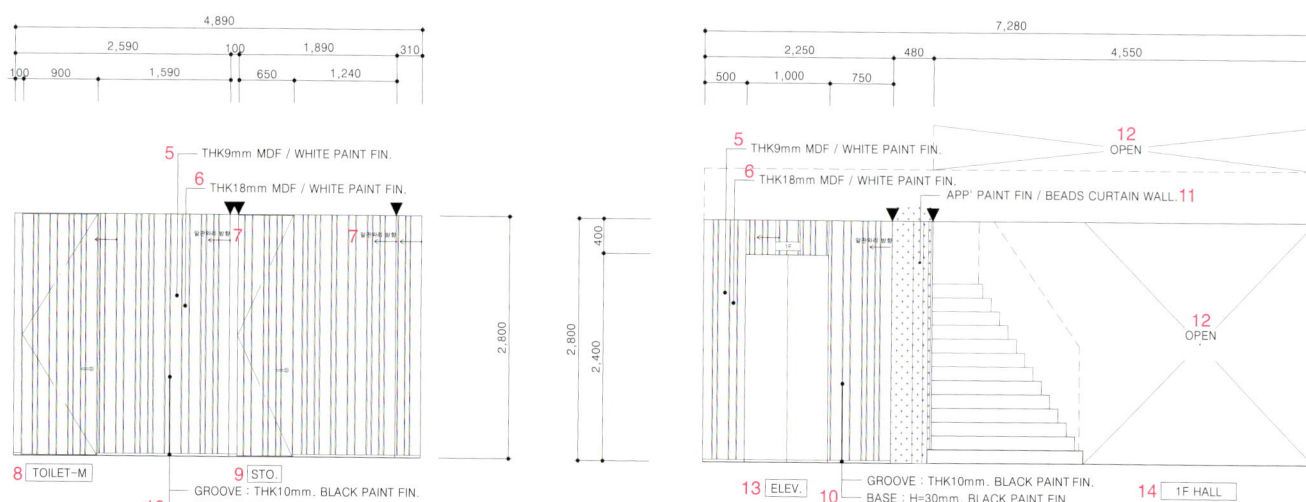

1층 로비 입면 G / 1st floor lobby elevation G

1층 로비 입면 H / 1st floor lobby elevation H

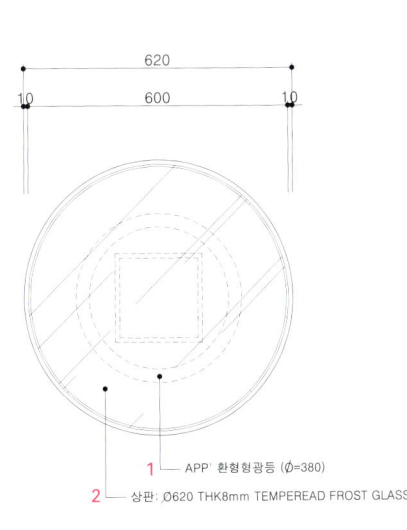
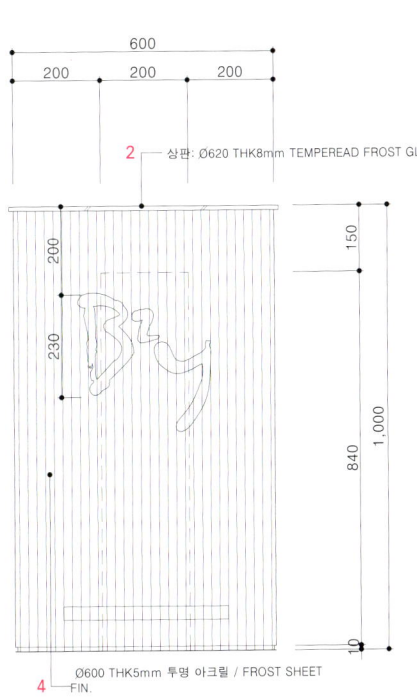
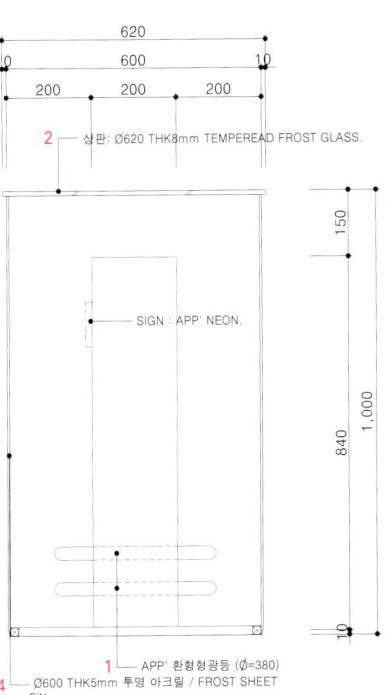

1층 로비 인포 데스크 상세 / 1st floor lobby info desk detail

1 BACKSIDE : FROST SHEET 1PLY / FRONT : LINE FROST SHEET, BLACK LINE SHEET 1PLY 2 BACKSIDE : FROST SHEET 1PLY / FRONT : BLACK LINE SHEET 1PLY / BACKSIDE : FROST SHEET 1PLY, FRONT : LINE FROST SHEET 1PLY / ALL BACKSIDE : FROST SHEET 1PLY 3 ALL GLASS - FROST SHEET 1PLY, LINE FROST SHEET, BLACK LINE SHEET 1PLY 4 OPEN

1 후면 : 서리 쉬트 1겹 / 전면 : 라인 서리 쉬트, 블랙 라인 쉬트 1겹 2 후면 : 서리 쉬트 1겹 / 전면 : 블랙 라인 쉬트 1겹 / 후면 : 서리 쉬트 1겹, 전면 : 라인 서리 쉬트 1겹 / 후면전체 : 서리 쉬트 1겹 3 모든 유리 - 서리 쉬트 1겹, 라인 서리 쉬트, 블랙 라인 쉬트 1겹 4 열림

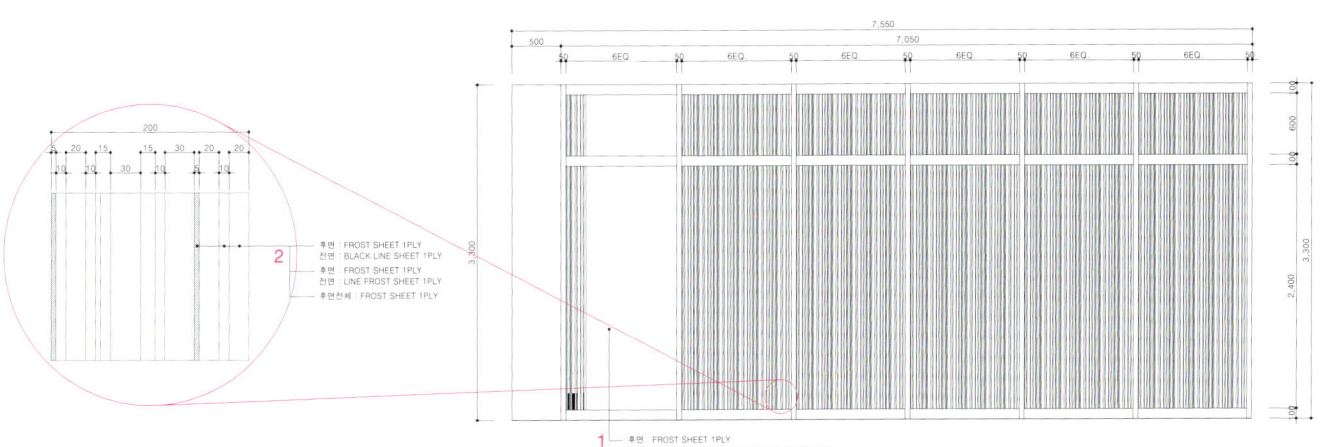

홀 벽 상세 J / hall wall detail J

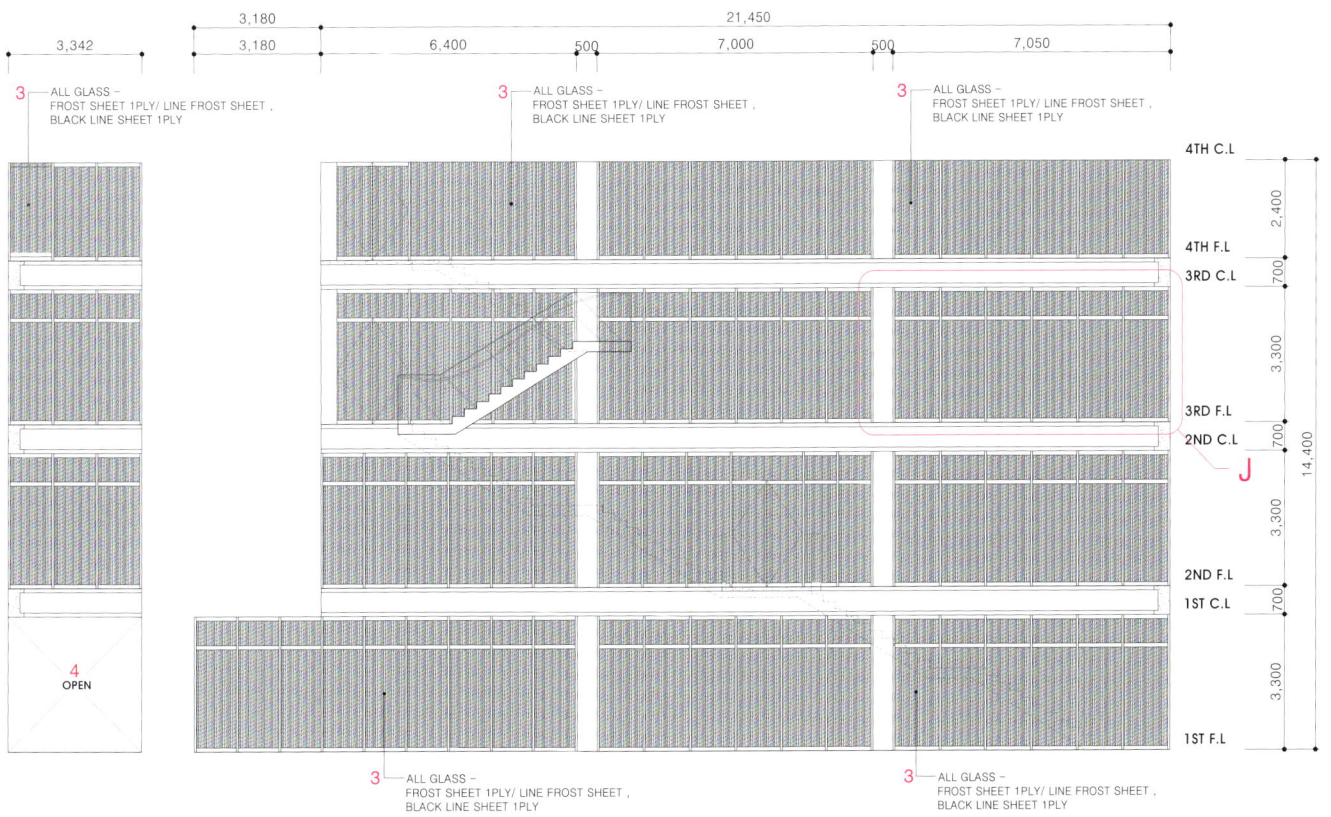

1층 홀 계단실 입면 I / 1st floor hall staircase elevation I

  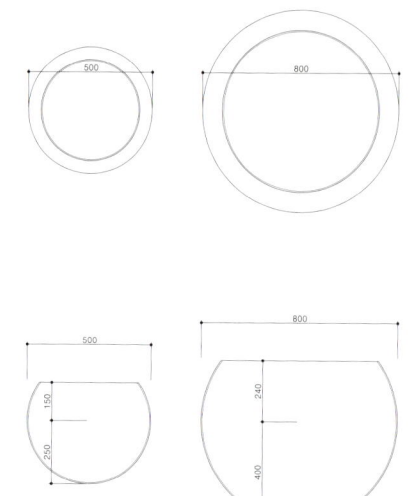

제작 조명 내부 아크릴 돔 상세 / design lighting inside acryl dome detail

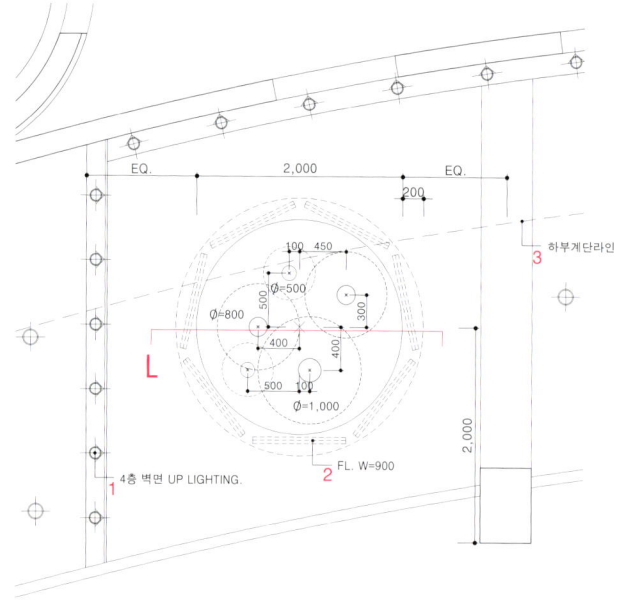

부분 천장 K / partial ceiling plan K

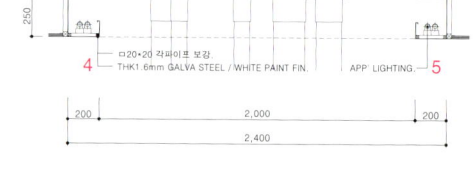

천장 단면 L / ceiling section L

1층 천장 키맵 / 1st floor ceiling key map

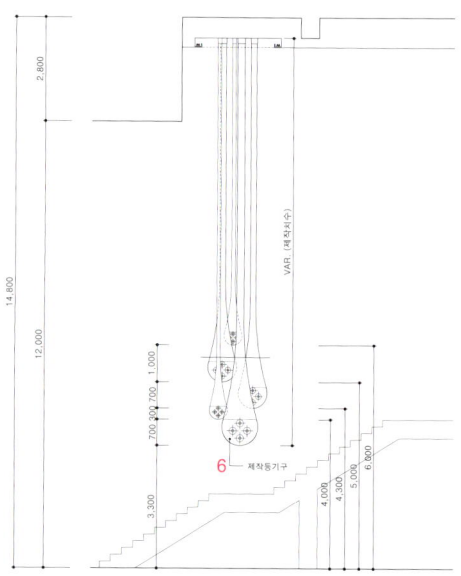

홀 제작 조명 입면 / hall design lighting elevation

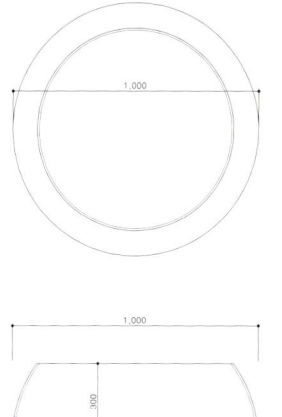
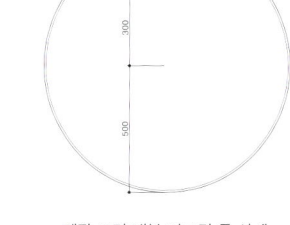

제작 조명 내부 아크릴 돔 상세
/ design lighting inside acryl dome detail

1 4TH FLOOR WALL UP LIGHTING 2 FL. W=900 3 LOWER PART STAIR LINE 4 □-20X20 SQUARED PIPE REINFORCE / T1.6 GALVA SHEET, WHITE PAINT FIN. 5 APP. LIGHTING 6 MAKING LIGHTING FIXTURES 7 APP. V.P FIN. 8 MAKING LIGHTING FIXTURES 5EA 9 Ø=2,000 10 APP. SPAN - FABRIC FIN. : MAKING LIGHTING FIXTURES BODY 11 LOBBY 12 WHITE MOSAIC TILE FIN. 13 TOILET - W 14 TOILET - M 15 APP. RED NEON / APP. BARRISOL FIN.(INSERT LIGHTING) 16 T3 STEEL PLATE 17 PERFORATING BOLT JOINT(FIELD NEGOTIATION) 18 APP. LEATHER FIN. / APP. FIXED RING 19 T3 STEEL PLATE : MAKING LIGHTING FIXTURES FIXED RING / APP. LEATHER FIN. : MAKING LIGHTING FIXTURES HEAD / T3 STEEL PLATE : MAKING LIGHTING FIXTURES FIXED FRAME 20 APP. BOLT JOINT / T3 STEEL PLATE : LIGHTING BOX CEILING PLATE

1 4층 벽면 업라이트 2 바닥. W=900 3 하부 계단 라인 4 □-20X20 각파이프 보강 / T1.6 갈바쉬트, 흰색 페인트 마감 5 지정 조명 6 제작 등기구 7 지정 도장 마감 8 제작 조명기구 5개 9 Ø=2,000 10 지정 스판 패브릭 마감 : 제작조명기구 바디 11 로비 12 흰색 모자이크 타일 마감 13 화장실(여) 14 화장실(남) 15 지정 레드 네온 / 지정 바리솔 마감(삽입 조명) 16 T3 철판 17 볼트 조인트 타공(현장협의) 18 지정 가죽 마감 / 지정 고정용 링 19 T3 철판 : 제작조명기구 고정링 / 지정 가죽 마감 : 제작조명기구 헤드 / T3 철판 : 제작조명기구 고정틀 20 지정 볼트 조인트 / T3 철판 : 등박스 천정면

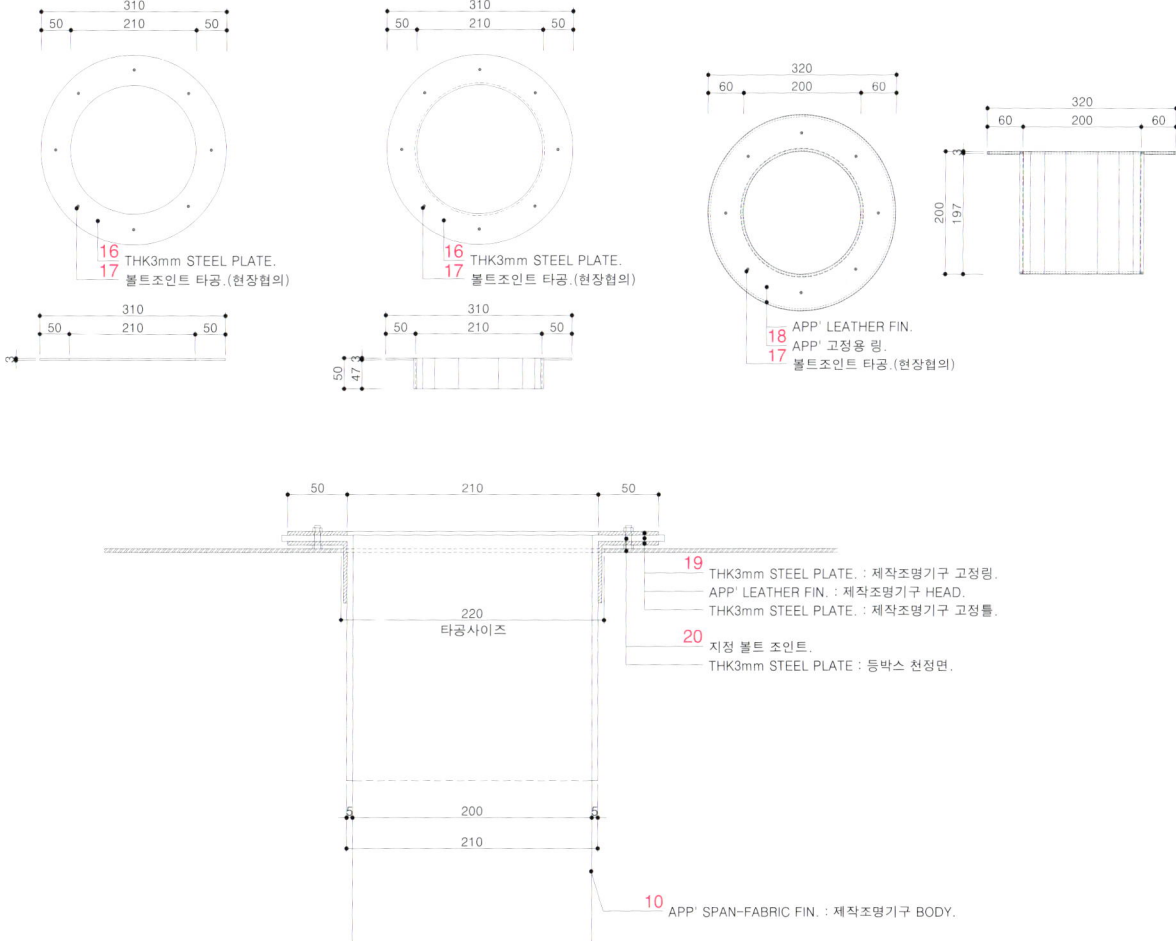

제작 조명 고정 상세 / design lighting fixing detail

1 SOFA : Ø=2,200 APP. LEATHER FIN.(WHITE) 2 STOOL : Ø=500 APP. LEATHER FIN.(WHITE) 3 APP. LEATHER FIN.(WHITE) 4 OPEN 5 TOILET - W 6 GROOVE : T10 BLACK PAINT FIN. / BASE : H=30MM, BLACK PAINT FIN. 7 T18 MDF, WHITE PAINT FIN. / T9 MDF, WHITE PAINT FIN. 8 HALOGEN SPOT. 9 T8 CLEAR TEMPERED GLASS(INSTALLING LOCKING SYSTEM) 10 INSIDE : APP. FABRIC FIN. 11 LIGHTING FIXTURES L5 - SPOT HALOGEN PERFORATED Ø95 12 APP. WIRE NET, H=280, INSTALLING WHITE CABLE TILE(RED COLOR RING SHAPE FLUORESCENT LAMP 8EA) 13 APP. WIRE NET, INSTALLING WHITE CABLE TILE 14 APP. WIRE NET, H=140, INSTALLING WHITE CABLE 15 APP. WIRE NET, H=280, INSTALLING WHITE CABLE TILE(RED COLOR RING SHAPE FLUORESCENT LAMP 14EA)

1 소파 : Ø=2,200 지정 가죽 마감(흰색) 2 스툴 : Ø=500 지정 가죽 마감(흰색) 3 지정 가죽 마감(흰색) 4 열림 5 화장실(여) 6 홈줄 : T10 블랙 페인트 마감 / 베이스 : H=30MM, 블랙 페인트 마감 7 T18 MDF, 흰색 페인트 마감 / T9 MDF, 흰색 페인트 마감 8 스팟 할로겐 조명 9 T8 투명강화유리(시건장치 설치) 10 내부 : 지정 패브릭 마감 11 조명기구 L5 – 스팟 할로겐 Ø95 타공 12 지정 철망, H=280, 흰색 케이블 타일 취부(레드컬러 환형형광등 8개) 13 지정 철망, 흰색 케이블 타일 취부 14 지정 철망, T, H=140, 흰색 케이블 취부 15 지정 철망, H=280, 흰색 케이블 타일 취부(레드 컬러 환형 형광등 14개)

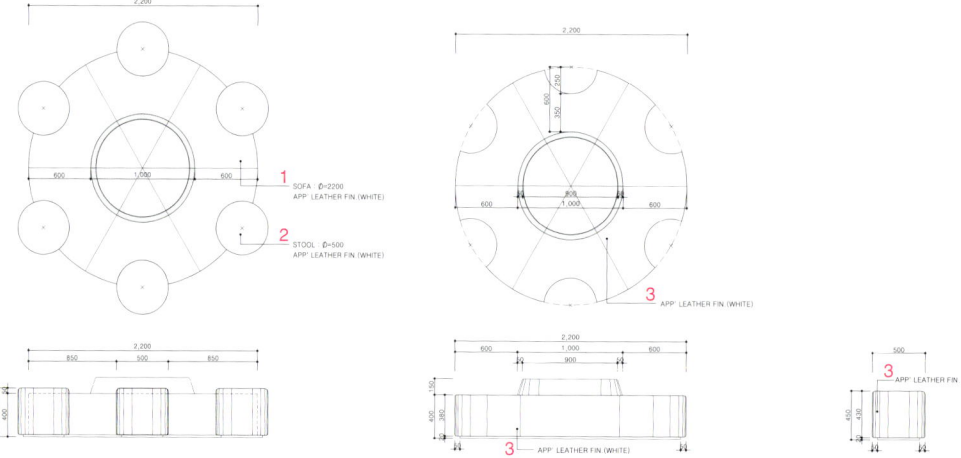

4층 라운지 제작 소파 상세 / 4th floor rounge design sofa detail

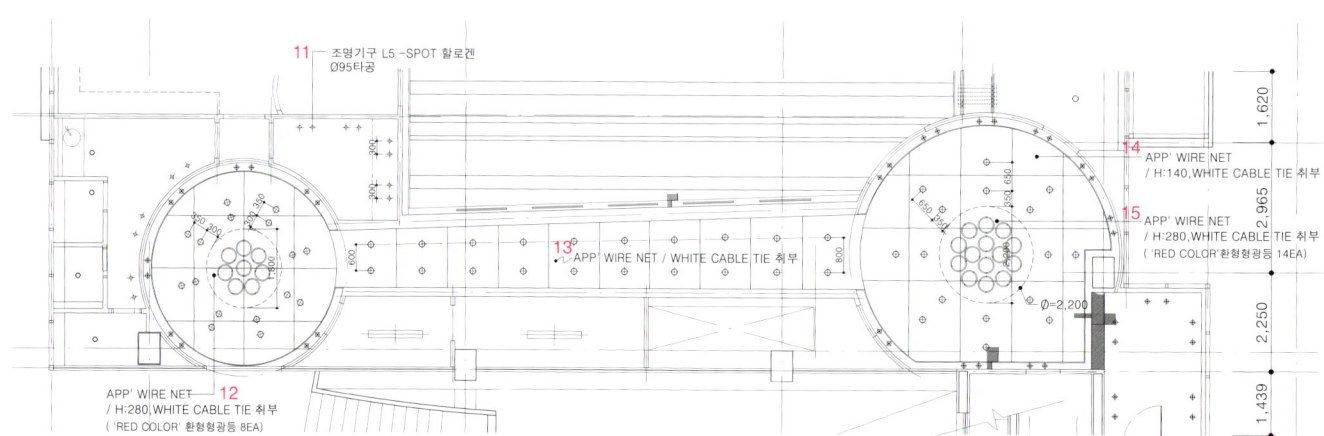

4층 부분 천장도 / 4th floor partial ceiling plan

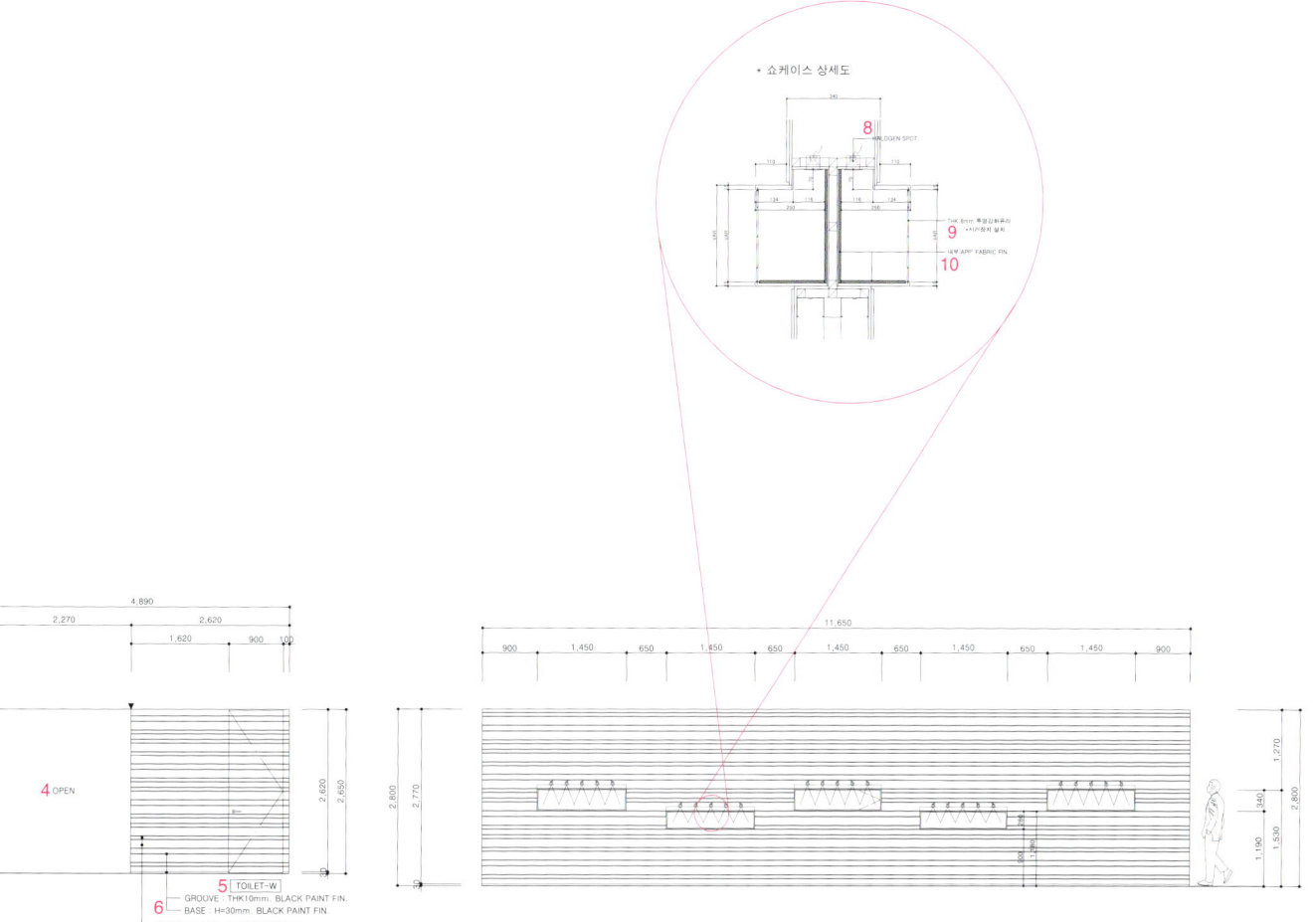

4층 엘리베이터 홀 입면 M / 4th floor elev. hall elevation M

4층 복도 입면 N / 4th floor corridor elevation N

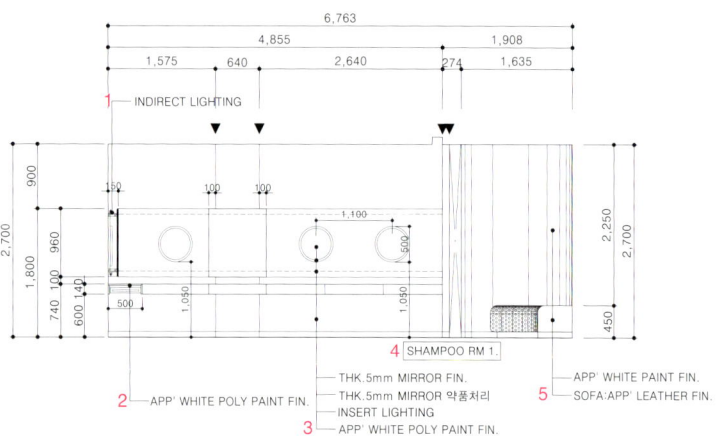

4층 미용실 입면 O / 4th floor beauty salon elevation O

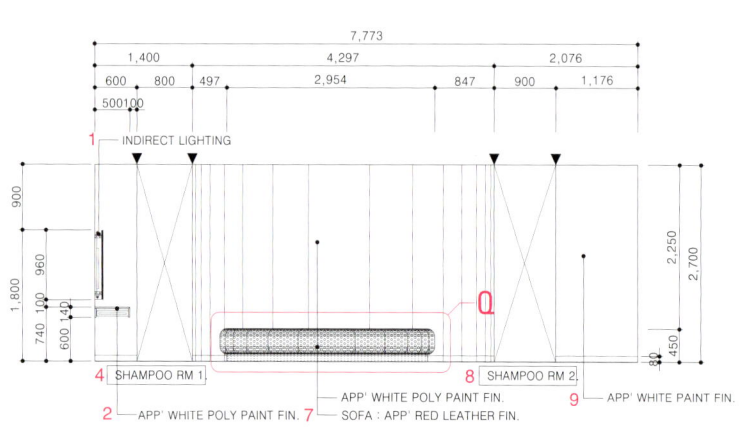

4층 미용실 입면 P / 4th floor beauty salon elevation P

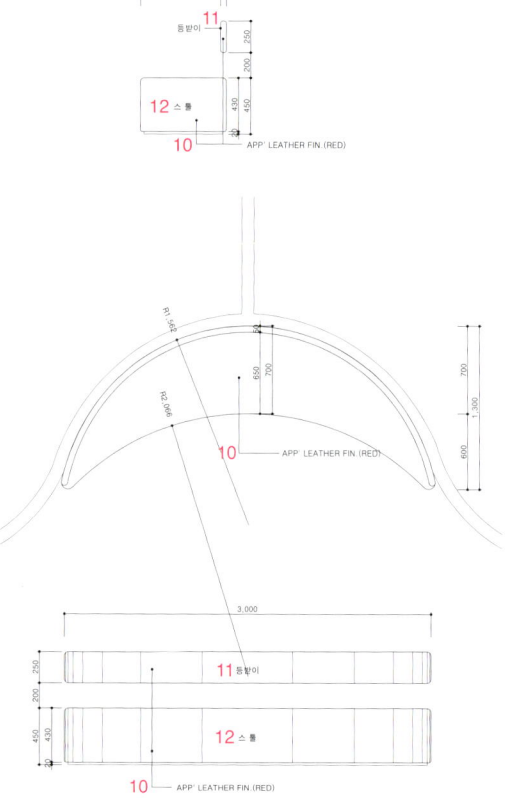

제작 소파 상세 Q / design sofa detail Q

1 INDIRECT LIGHTING 2 APP. WHITE POLY PAINT FIN. 3 T5 MIRROR FIN. / T5 MIRROR CHEMICALS TREATMENT / INSERT LIGHTING / APP. WHITE POLY PAINT FIN. 4 SHAMPOO RM1 5 APP. WHITE PAINT FIN. / SOFA : APP. LEATHER FIN. 6 BASE : SUS H/L FIN. 7 APP. WHITE POLY PAINT FIN. / SOFA : APP. RED LEATHER FIN. 8 SHAMPOO RM2 9 APP. WHITE FIN. 10 APP. LEATHER FIN.(RED) 11 BACKBOARD 12 STOOL 13 UPPER & LOWER PART LIGHTING EMBEDDING / T5 MIRROR FIN. 14 WHITE MOSAIC TILE FIN. 15 WHITE PAINT FIN. / GROOVE : BLACK PAINT FIN. / BASE : SUS H/R FIN. 16 T10 MIRROR FIN. 17 T5 MIRROR FIN. 18 INSTALLING PAPER TOWEL BOX 19 MAKING WASHBOWL / T10 MIRROR FIN. 20 OPEN SIZE 280X730 21 WHITE PAINT FIN / GROOVE : BLACK PAINT FIN. 22 LAUNDRY

1 간접 조명 2 지정 흰색 폴리페인트 마감 3 T5 거울 마감 / T5 거울 약품처리 / 삽입 조명 / 지정 흰색 폴리페인트 마감 4 샴푸실1 5 지정 흰색 페인트 마감 / 소파 : 지정 가죽 마감 6 베이스 : 스텐레스 헤어라인 마감 7 지정 흰색 폴리페인트 마감 8 샴푸실 2 9 지정 흰색 마감 10 지정 가죽 마감(레드) 11 등받이 12 스툴 13 상하부 라이팅 매입 / T5 거울 마감 14 흰색 모자이크 타일 마감 15 흰색 페인트 마감 / 홈줄: 블랙 페인트 마감 / 베이스 : 스테인레스 헤어라인 마감 16 T10 거울 마감 17 T5 거울 마감 18 종이 수건 박스 설치 19 제작세면볼 / T10 거울 마감 20 오픈 사이즈 280X730 21 흰색 페인트 마감 / 홈줄: 블랙 페인트 마감 22 세탁실

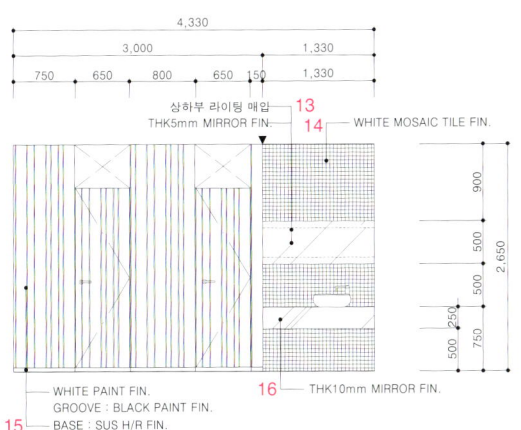

4층 화장실 입면 R / 2th floor toilet elevation R

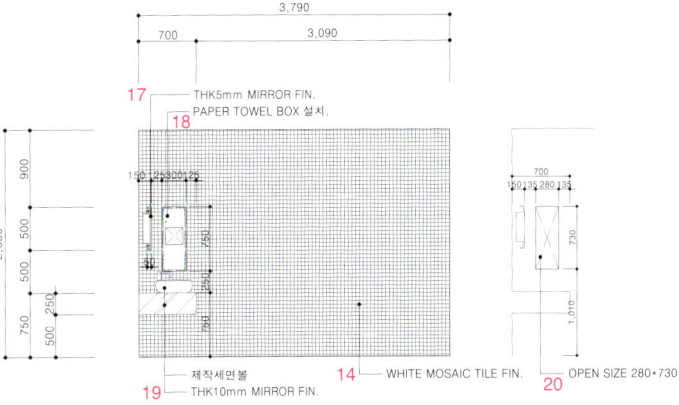

4층 화장실 입면 S / 2th floor toilet elevation S

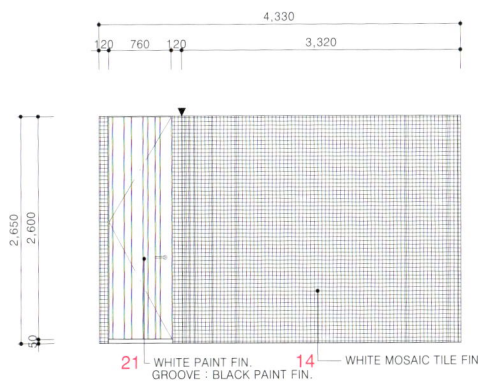

4층 화장실 입면 T / 2th floor toilet elevation T

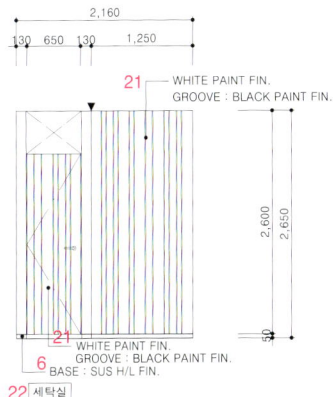

4층 화장실 입면 U / 2th floor toilet elevation U

YOUNGONE HEAD OFFICE REMODELING

(주)다원디자인 | 조서윤   DAWON INTERIOR DESIGN | Fay Cho

국내 1위의 아웃도어 브랜드인 (주)영원무역은 아웃도어 제품을 주력으로 생산하는 대표적인 패션회사로서 이번 프로젝트를 시작함에 있어서 던진 화두는 "What do you want?" 이다.
패션회사로써의 이미지, 아이덴티티를 구축하기 위해서 우리는 위의 물음에 다음과 같은 결론을 도출시킴으로써 프로젝트의 시발점으로 삼았다.

1. Exterior - I want to go there.
2. Office - I want to work there.
3. Show room - I want to show it.
4. Lobby - I want to know more.
5. Cafeteria - I want to be free there.

The leading domestic outdoor brand Youngone is a representative fashion company which mainly produces outdoor goods. As we commenced this project for the company, we focused on the very basic question "what do you want?". In order to establish images and an identity of the fashion company, we started the project with drawing the conclusions as follows.

1. Exterior - I want to go there.
2. Office - I want to work there.
3. Show room - I want to show it.
4. Lobby - I want to know more.
5. Cafeteria - I want to be free there.

이렇게 크게 5개의 공간, 5개의 테마는 각 공간의 디자인 모티브로 활용되며 전체 공간을 하나의 컨셉 "Mecca of modern outdoor life" 라는 영원무역의 아이엔티티를 표현 하고자 하였다.

Those five spaces represent "Mecca of modern outdoor life", the identity of Yongone Trading through five themes which were used as design motives for the devided spaces.

위치 서울시 중구 만리동2가 171번지
규모 지하 4층 ~ 지상 8층
건축면적 888.11㎡
설계팀 (주)다원디자인 / 조서윤, 김주상
시공 (주)다원디자인
마감 바닥 – 실버후지, 우드 플로링, 프로텍스, 카펫 타일, 러그, 벽 – 아줄 그레이, 비닐 페인트, 유리, 청동 바이브레이션, 새틴 바이브레이션, 패브릭, 벽지, 천장 – 비닐 페인트, 바리솔, 루버
사진 (주)다원디자인 제공

Location 171, Malli-dong 2-ga, Jung-gu, Seoul, Korea
Building scope B4 ~ 8F
Building area 888.11㎡
Design team Dawon Design Co., Ltd. / Fay Cho, Kim Ju Sang
Construction Dawon Design Co., Ltd.
Finishing Floor - Silver fuji, Wood flooring, flotex, Carpet tile, Rug, Wall - Azul grey, Vunyl paint, Glass, Bronze vibration, Satin vubration, Fabric, Wall covering, Ceiling - Vinyl patint, Barrisol, Louver
Photos offer Dawon Design Co., Ltd.

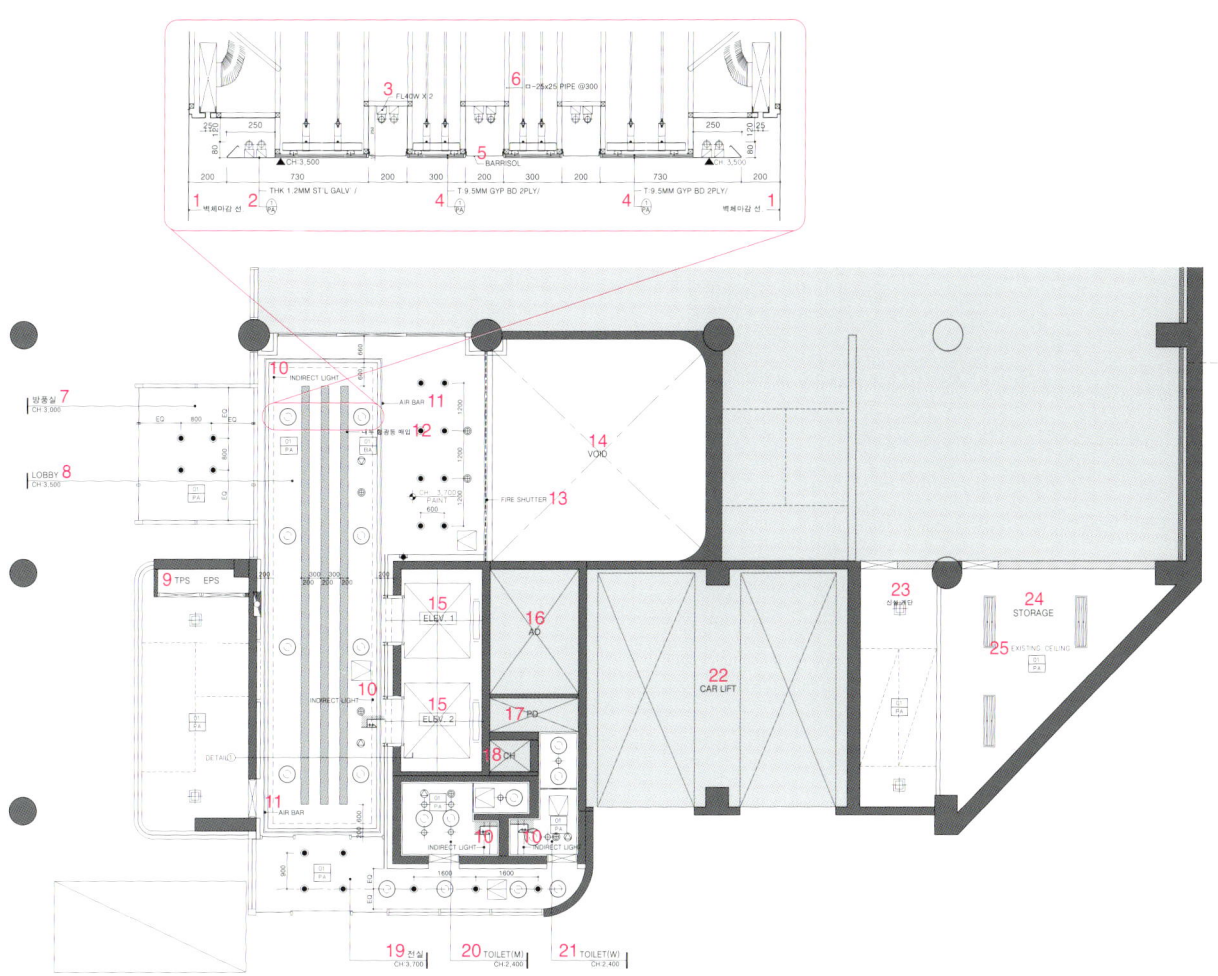

1층(입구 홀) 천장도 / 1st(entrance hall) ceiling plan

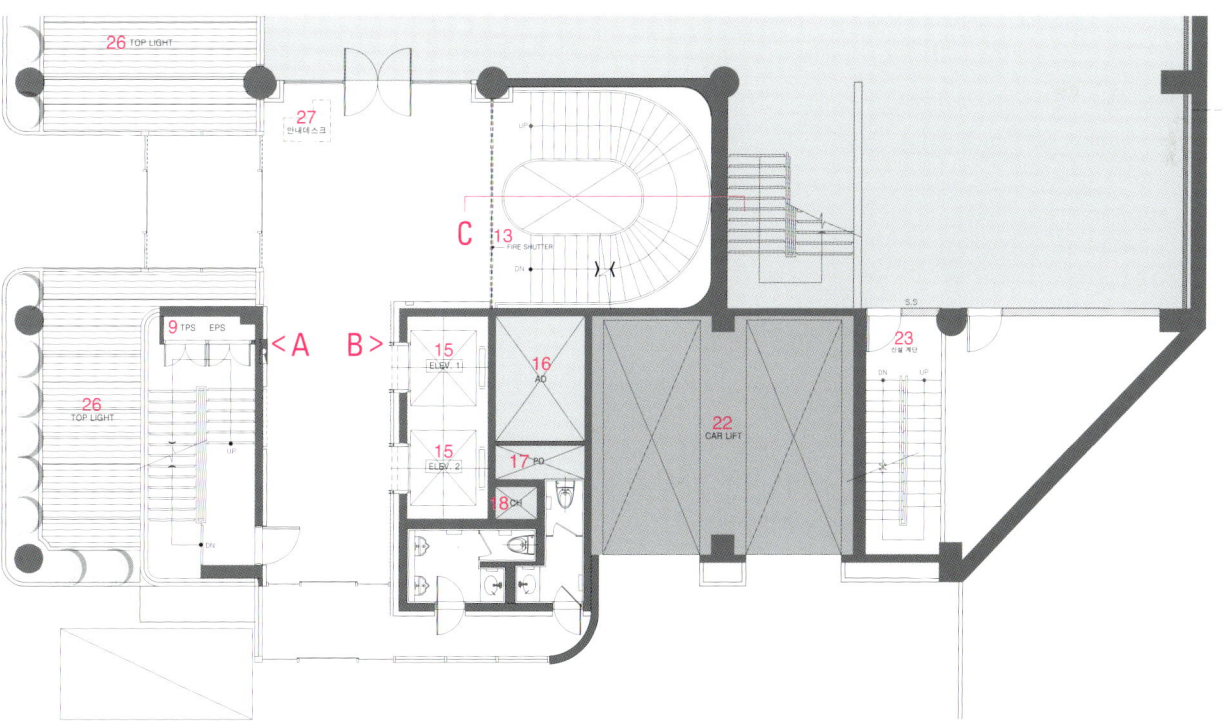

1층(입구 홀) 평면도 / 1st(entracne hall) floor plan

1 WALL FINISHING LINE 2 T1.2 ST'L GALV. 3 FL - 40WX2 4 T9.5 GYP BD 2PLY 5 BARRISOL 6 □ - 25X25 PIPE @300 7 AIR ROOM 8 LOBBY 9 TPS / EPS 10 INDIRECT LIGHT 11 AIR BAR 12 INSIDE EMBEDDED FLUORESCENT LAMP 13 FIRE SHUTTER 14 VOID 15 ELEV. 16 AD 17 PD 18 CH 19 FRONT ROOM 20 TOILET(M) 21 TOILET(W) 22 CAR LIFT 23 STAIR 24 STORAGE 25 EXISTING CEILING 26 TOP LIGHT 27 INFO.DESK

1 벽체 마감선 2 T1.2 스틸 갈바 3 FL - 40WX2 4 T9.5 석고보드 2겹 5 바리솔 6 □ - 25X25 파이프 @300 7 방풍실 8 로비 9 TPS / EPS 10 간접 조명 11 에어 바 12 내부 형광등 매입 13 방화 셔터 14 보이드 15 엘리베이터 16 AD 17 PD 18 CH 19 전실 20 남자 화장실 21 여자 화장실 22 주차 리프트 23 신설계단 24 창고 25 기존 천장 26 탑 라이트 27 안내데스크

YOUNGONE HEAD OFFICE REMODELING | 영원무역 본사

1 FIRE SHUTTER 2 BUTT JOINT 3 T5 REVEAL, T50 BASE 4 200(W)X 250(H)X140(D) 5 LIGHTING(LED) 6 200(W)X15(H)X30(D) 7 EXISTING WINDOW 8 EXISTING DOOR 9 STAIR CASE 10 REPAINT 11 INDIRECT LIGHTING 12 EMBEDDED PDP(60") 13 FOOT LIGHTING (LED) 14 EXISTING FEC. 15 AUTO DOOR 16 FRAME 17 AIR ROOM 18 OPEN TO STAIR CASE 19 STAIR 20 HAND RAIL 21 FLOOR SIGN 22 ELEV. 23 T15 CRYSTAL GLASS 24 LIGHTING BOX(LED) 25 OPEN TO CORRIDOR 26 T9.5 GYP. BOARD 2PLY, METALLIC PAINTING FIN. 27 GROOVING 28 Ø60 STEEL PIPE, APP. V.P FIN. 29 ㅁ - 50X50 PIPE 30 T1.2 ST'L GALV. 31 T30 / T45 DRY CEMENT MORTAR, T5 CEMENT PASTE

1 방화셔터 2 맞대이음 3 T5 창, T50 베이스 4 200(W)X250(H)X140(D) 5 조명(LED) 6 200(W)X15(H)X30(D) 7 기존 창호 8 기존 문 9 계단실 10 재도장 11 간접조명 12 PDP(60") 매입 13 바닥 조명(LED) 14 기존 FEC. 15 자동문 16 프레임 17 방풍실 18 계단실 열린공간 19 계단 20 난간 21 층별 사인 22 엘리베이터 23 T15 크리스탈 유리 24 조명 박식(LED) 25 복도 열린공간 26 T9.5 석고보드 2겹, 메탈릭 도장 마감 27 홈파기 28 Ø60 스틸 파이프, 지정 도장 마감 29 ㅁ - 50X50 파이프 30 T1.2 스틸 갈바 31 T30 / T45 건시멘트 몰탈, T5 시멘트페이스트

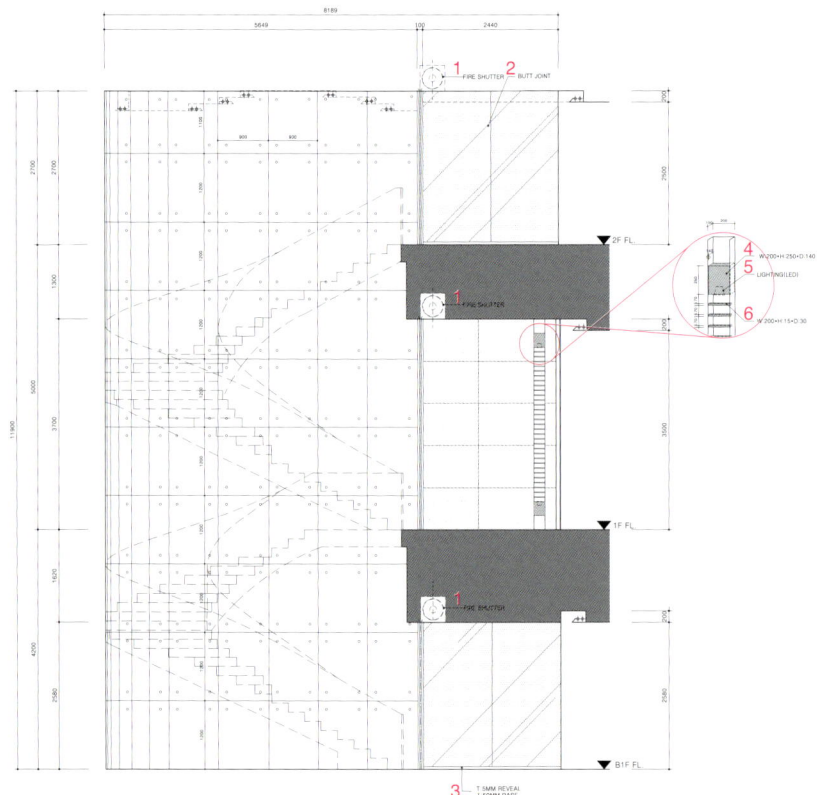

계단실 단면 C / staircase section C

1층 입구 홀 입면 A / 1F entrance hall elevation A

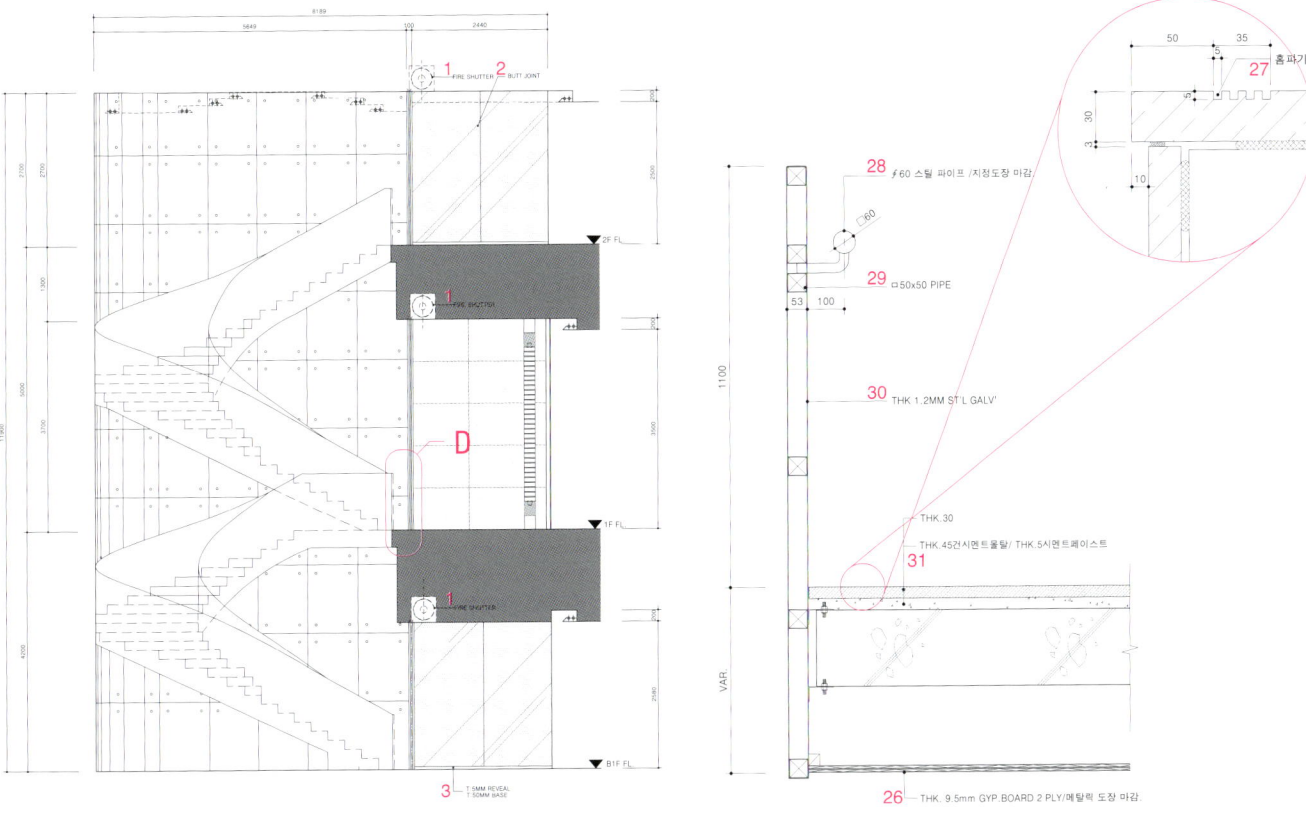

계단실 단면 C' / staircase section C'

난간 상세 D / handrail detail D

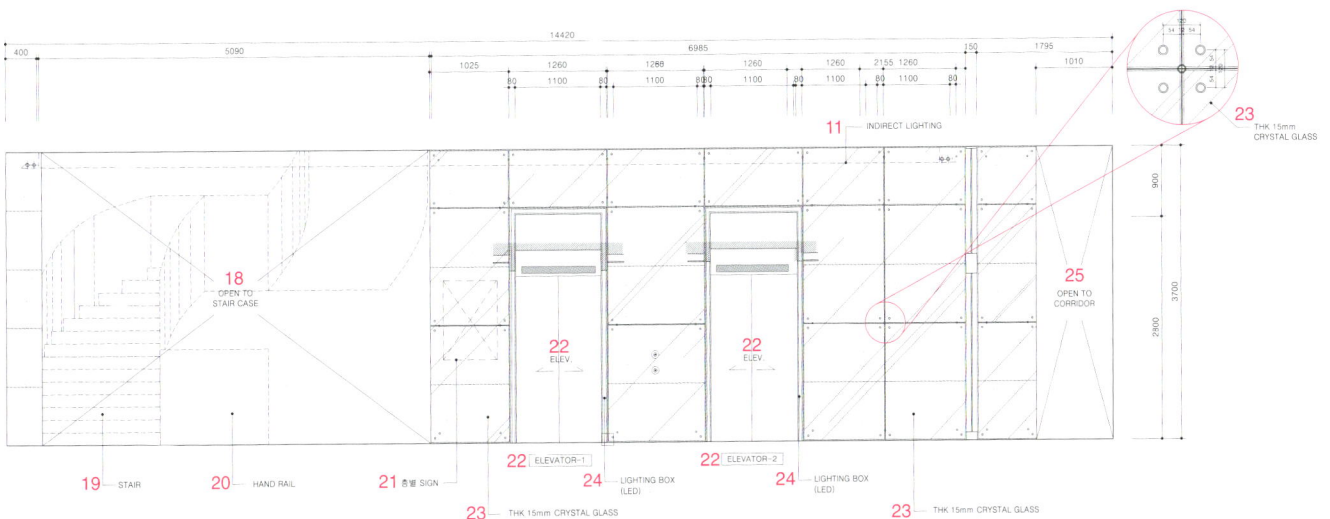

1층 입구 홀 입면 B / 1F entrance hall elevation B

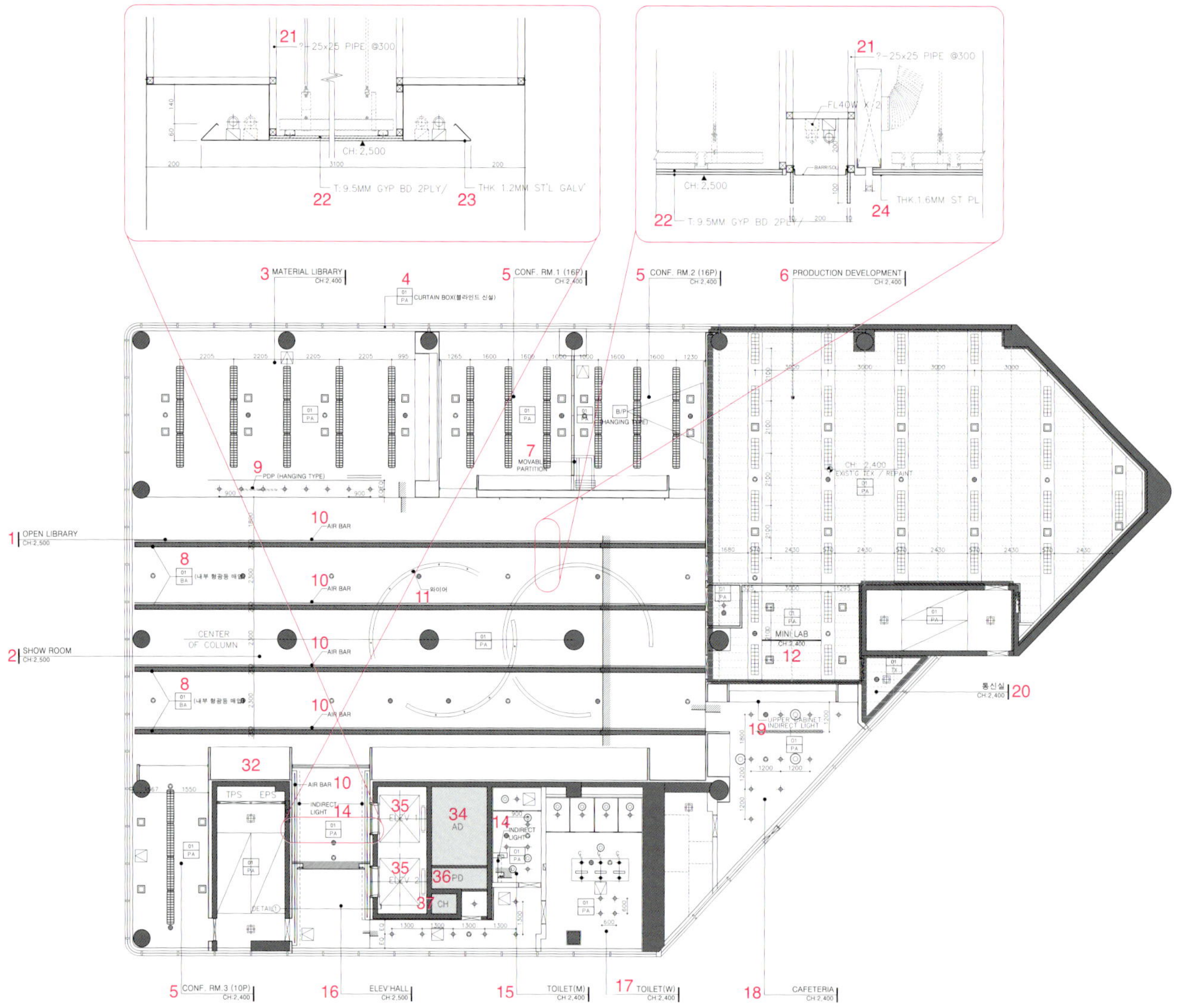

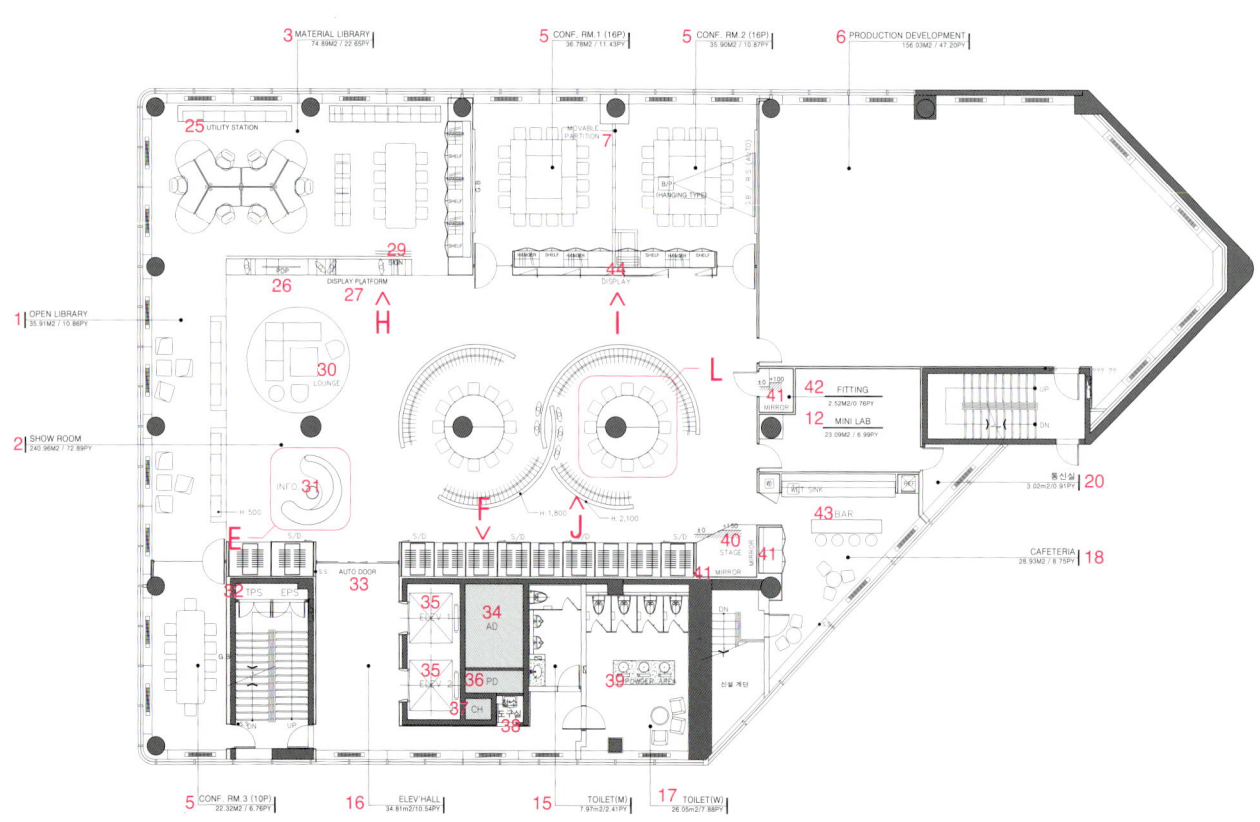

3층(쇼룸) 천장도 / 3rd (show room) ceiling plan

3층(쇼룸) 평면도 / 3rd (show room) floor plan

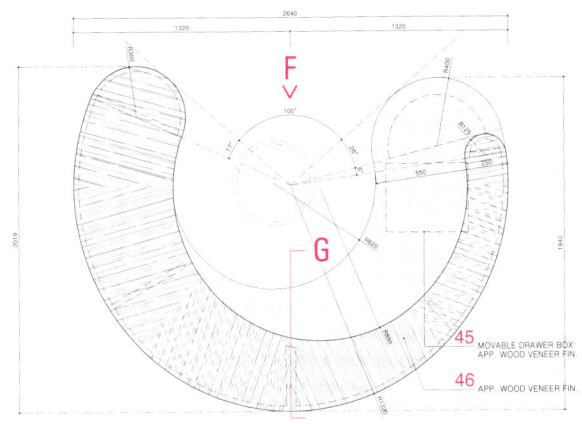

3층 쇼룸 인포데스크 평면 E / 3F show room info desk plan E

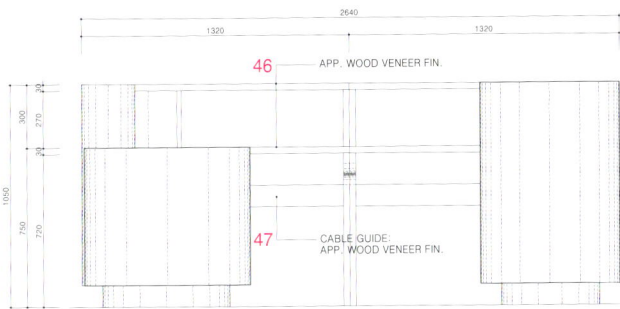
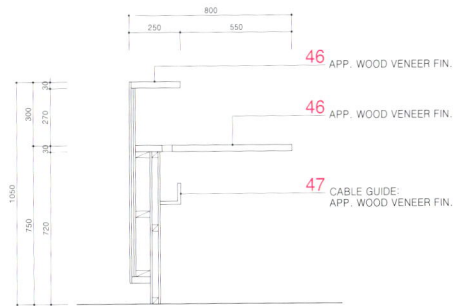

인포데스크 입면 F / info desk elevation F

인포데스크 단면 G / info desk section G

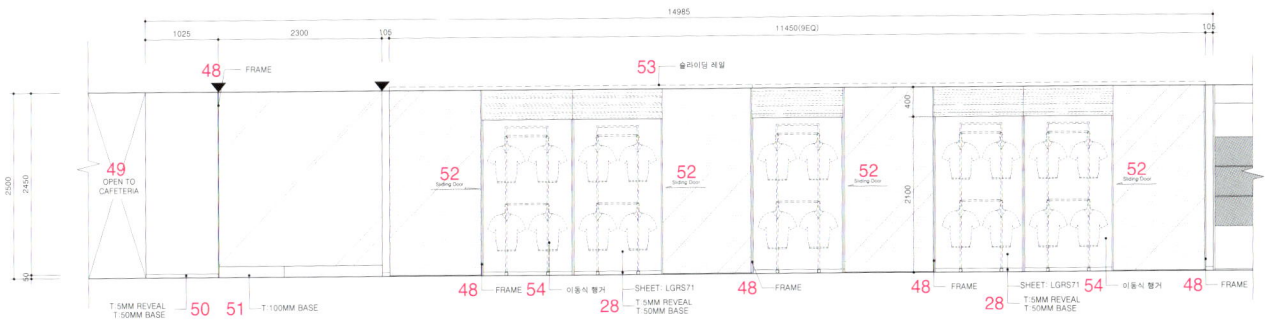

쇼룸(닫힘) 입면 F / show room(door closed) elevation F

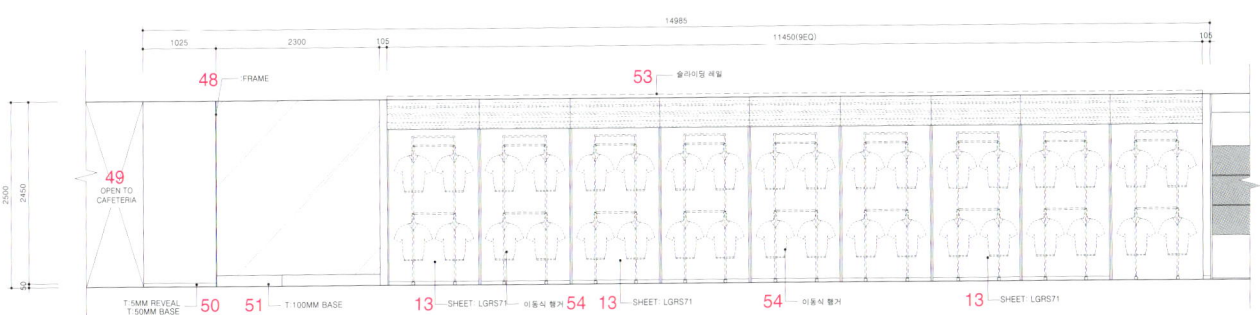

쇼룸(열림) 입면 F' / show room(door opened) elevation F'

1 OPEN LIBRARY 2 SHOW ROOM 3 MATERIAL LIBRARY 4 CURTAIN BOX(NEWLY-BLIND) 5 CONF. RM.(16P) 6 PRODUCTION DEVELOPMENT 7 MOVABLE PARTITION 8 INSIDE EMBEDDED FLUORESCENT LAMP 9 PDP(HANGING TYPE) 10 AIR BAR 11 WIRE 12 MINI LAB 13 SHEET : LGRS71 14 INDIRECT LIGHT 15 TOILET(M) 16 ELEV. HALL 17 TOILET(W) 18 CAFERTRIA 19 UPPER CABINET : INDIRECT LIGHT 20 CORRESPONDENCE ROOM 21 □ - 25X25 PIPE @300 22 T9.5 GYP BD 2PLY 23 T1.2 ST'L GALV 24 T1.6 ST PL 25 UTILITY STATION 26 PDP 27 DISPLAY PLATFORM 28 SHEET : LGRS71 / T5 REVEAL, T50 BASE 29 SIGN 30 LOUNGE 31 INFO. 32 TPS / EPS 33 AUTO DOOR 34 AD 35 ELEV. 36 PD 37 CH 38 DUSTING THINGS ROOM 39 POWDER AREA 40 STAGE 41 MIRROR 42 FITTING 43 BAR 44 DISPLAY 45 MOVABLE DRAWER BOX : APP. WOOD VENEER FIN. 46 APP. WOOD VENEER FIN. 47 CABLE GUIDE : APP. WOOD VENEER FIN. 48 FRAME 49 OPEN TO CAFETERIA 50 T5 REVEAL, T50 BASE 51 T100 BASE 52 SLIDING DOOR 53 SLIDING RAIL 54 MOVABLE HANGER

1 서고 열린공간 2 쇼룸 3 소재 자료실 4 커튼 박스(블라인드 신설) 5 회의실(16인) 6 상품개발 7 이동식 파티션 8 내부 형광등 매입 9 PDP(벽걸이) 10 에어 바 11 와이어 12 미니 랩 13 시트 : LGRS71 14 간접 조명 15 남자 화장실 16 엘리베이터 홀 17 여자 화장실 18 카페테리아 19 상부 캐비닛 : 간접 조명 20 통신실 21 □ - 25X25 파이프 @300 22 T9.5 석고보드 2겹 23 T1.2 스틸 갈바 24 T1.6 스틸 판 25 유틸리티 스테이션 26 PDP 27 디스플레이 플랫폼 28 시트 : LGRS71 / T5 창틀, T50 베이스 29 사인 30 라운지 31 인포메이션 32 TPS / EPS 33 자동문 34 AD 35 엘리베이터 36 PD 37 CH 38 청소 도구실 39 메이크업실 40 창고 41 거울 42 피팅실 43 바 44 디스플레이 45 이동식 선반박스 : 지정 무늬목 마감 46 지정 무늬목 마감 47 케이블 가이드 : 지정 무늬목 마감 48 틀 49 카페테리아 열린공간 50 T5 창틀, T50 베이스 51 T100 베이스 52 미닫이 문 53 슬라이딩 레일 54 이동식 행거

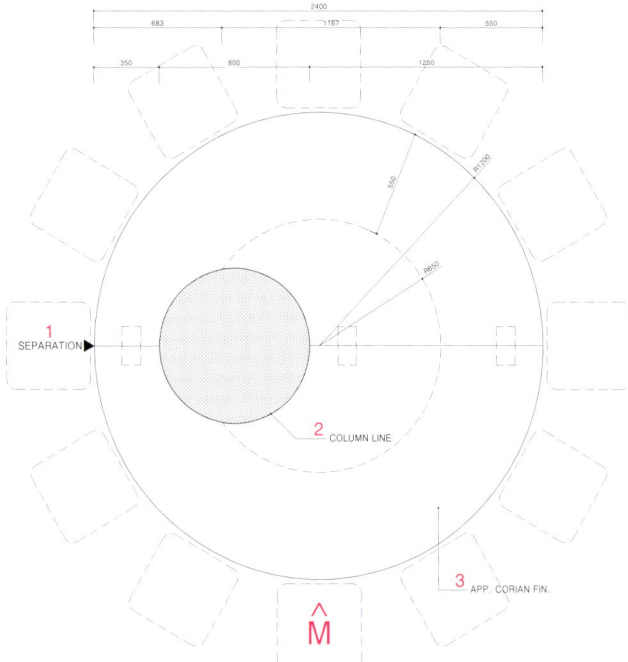

쇼룸 테이블 평면 L / show room table plan L

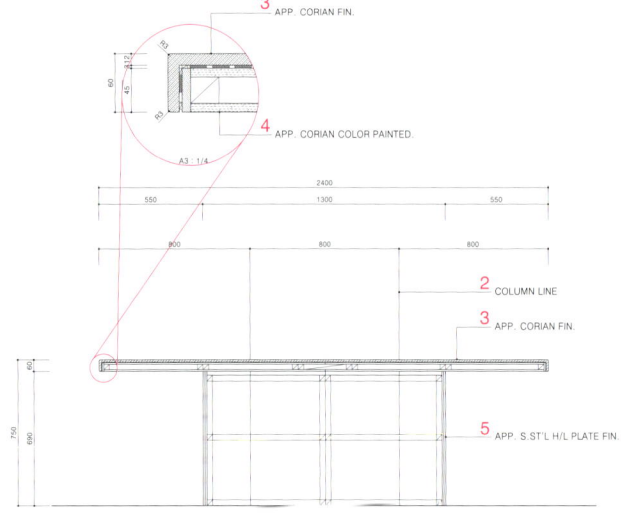

테이블 단면 N / table section N

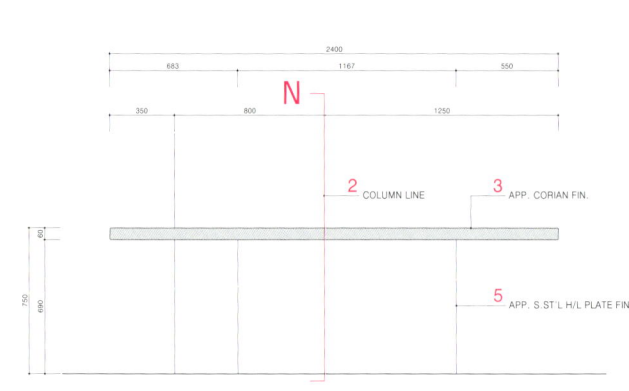

테이블 입면 M / table elevation M

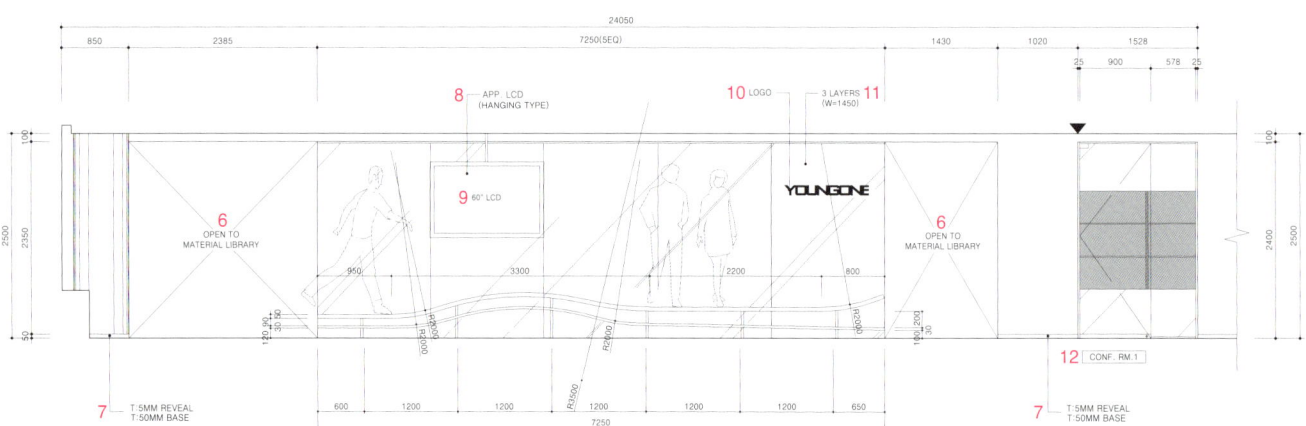

쇼룸 입면 H / show room elevation H

1 SEPARATION 2 COLUMN LINE 3 APP. CORIAN FIN. 4 APP. CORIAN COLOR PAINTED 5 APP. S.ST'L H/L PLATE FIN. 6 OPEN TO MATERIAL LIBRARY 7 T5 REVEAL, T50 BASE 8 APP. LCD (HANGING TYPE) 9 60" LCD 10 LOGO 11 3 LAYERS(W=1,450) 12 CONF. RM 13 COLUMN 14 OPEN 15 WIRE 16 CENTER OF WALL 17 DISPLAY 18 T1.2 S. STEEL(SUPER MIRROR)

1 분리대 2 기둥라인 3 지정 코리안 마감 4 지정 코리안 컬러 페인트 5 지정 스테인레스 헤어라인 판 마감 6 소재 자료실 열린공간 7 T5 창틀, T50 베이스 8 지정 LCD(벽걸이) 9 60" LCD 10 로고 11 3 층(W=1,450) 12 회의실 13 기둥 14 열림 15 와이어 16 벽 중심 17 디스플레이 18 T1.2 스테인레스 스틸(슈퍼미러)

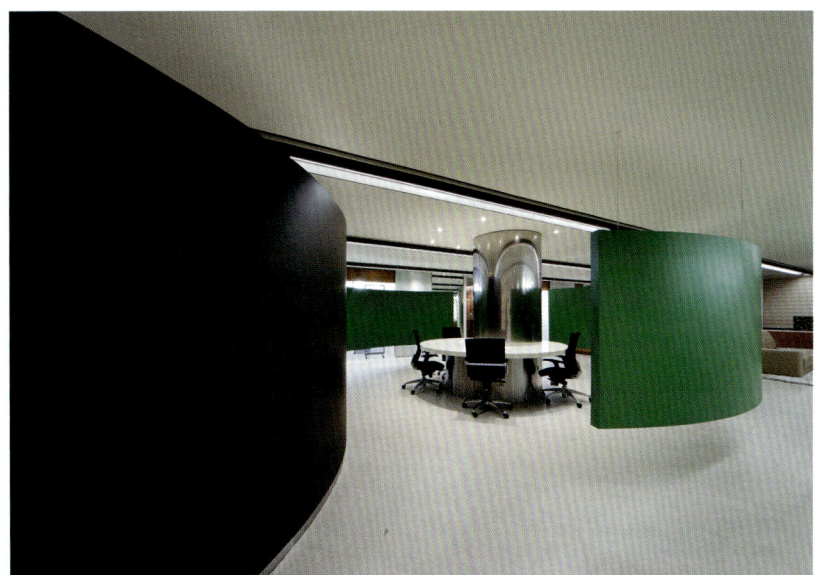

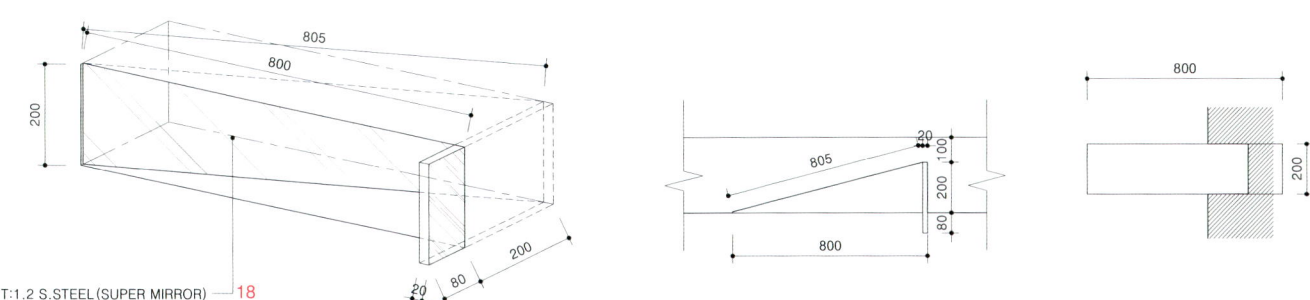

디스플레이 벽 상세 K / display wall detail K

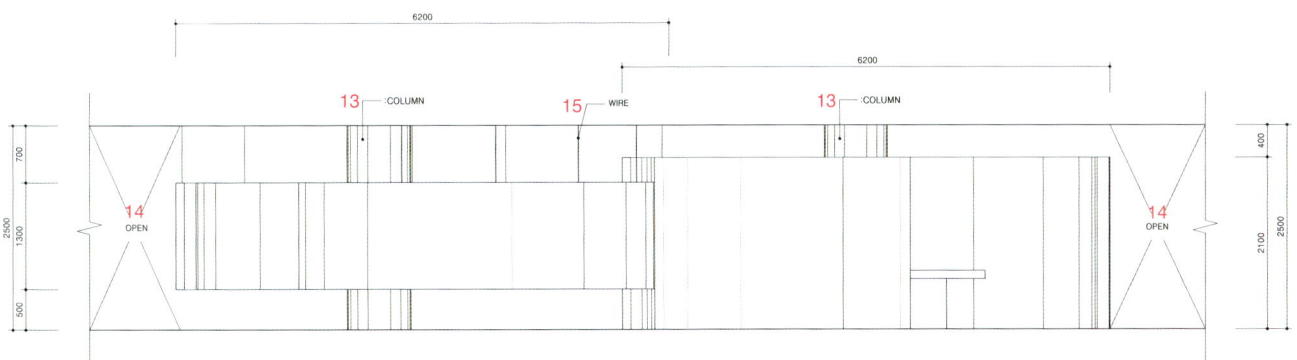

쇼룸 입면 J / show room elevation J

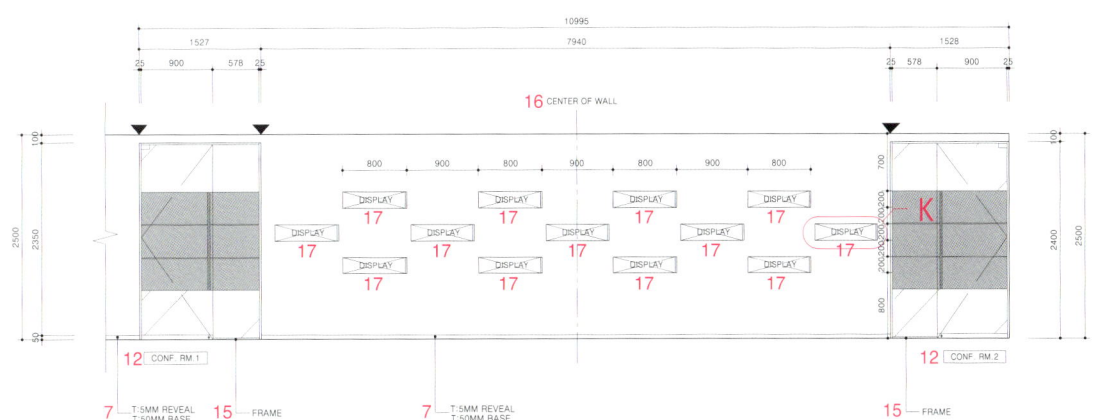

쇼룸 입면 I / show room elevation I

GREY GROUP

㈜다원디자인 | 조서윤 DAWON INTERIOR DESIGN | Fay Cho

클라이언트의 방문이 잦은 사무실은 그들의 환경만 보고도 얼마나 창조적인 조직인가를 느낄 수 있어야 한다. 광고 회사인 그레이월드 와이드 코리아를 디자인함에 있어서 공간에 담겨질 이야기, 즉 이곳을 사용하게 될 직원들을 우선으로 고려하고 그들의 사무공간에서 이뤄지는 작업들과 일상패턴을 파악하고 그에 따르는 디자인을 적용시키고자 하였다. 광고회사라는 특성상 끊임없는 아이디어 회의는 물론 일상생활 속에서도 그들은 겉으로 드러나는 것 외에도 더 많은 생각과 고민을 하고 있을 것이라 예상할 수 있다. 즉, 이들의 공간은 치열한 전쟁터와 다르지 않으므로 일반적인 사무공간에서 탈피하여 직업적 긍지를 가져다 줄 수 있는 곳이 되어야 했다. 이에 디자이너는 그들의 일상을 살표보던 중, 가장 흥미진진한 순간이 광고주가 방문하는 날이라는 것을 알게 되었다. 따라서 그 흥미진진한 날의 연속성이 이곳의 디자인 컨셉이 되었다.

첫인상을 좌우하는 2층 입구는 상호가 선명 새겨진 이미지월이 세워져 있으며 이어지는 리셉션에서는 오렌지 컬러와 천장에서의 스포트 라이트를 통해 공간감을 살렸다. 크레이브팀이 이용하는 3층은 자유로운 분위기의 사무실로 구성하여 창작의 자유를 마음껏 펼칠 수 있도록 하였다. 역동적인 디자인이 곳곳에 적용된 이곳은 연속적인 흐름을 가지면서도 기능에 따라 적절한 변화를 주며 함께 일하기 위한 즐거운 환경을 가져다 준다.

Modigliani is a brunch cafe that took over the space of a former fashion shop and it is composed of a 1st basement for small meetings or parties and a 1st floor for a brunch and dining cafe. Under the name of Italian artist, Modigliani, who was famous for portraits of women, they set the design direction to create a space that can have an atmosphere of a gallery or that provides insight to people who dream to imagine what Modigliani's atelier was like. Plants, moss and trees were drawn into the building to create a mood of having an outdoor meal and naturalism by objets helps to make the space a comfortable place.

The project focused on reduction rather than addition and the place was cleared in a way to emphasize physical properties and shapes of things.

위치 서울시 강남구 신사동 600-1 2~3층
용도 사무공간
면적 953㎡
설계팀 ㈜다원디자인 / 조서윤, 남용식, 김재환
시공 ㈜다원디자인
마감 바닥 – 세라믹 타일, 카펫 타일, 우드 플로링, 벽 – 도장, 패브릭, 강판, 천장 – 노출 천장, 비닐페인트, 스트레치 필름
사진 ㈜다원디자인 제공

Location 2~3F, 600-1, Sinsa-dong, Gangnam-gu, Seoul, Korea
Use Office
Area 953㎡
Design team Dawon Design Co., Ltd. / Fay Cho, Nam Yong Sik, Kim Jae Hwan
Construction Dawon Design Co., Ltd.
Finishing Floor - Ceramic tile, Carpet tile, Wood flooring, Wall - V.P, Fabric, Steel plate, Ceiling - Open ceiling, Vinyl paint, Stretch film
Photos offer Dawon Design Co., Ltd.

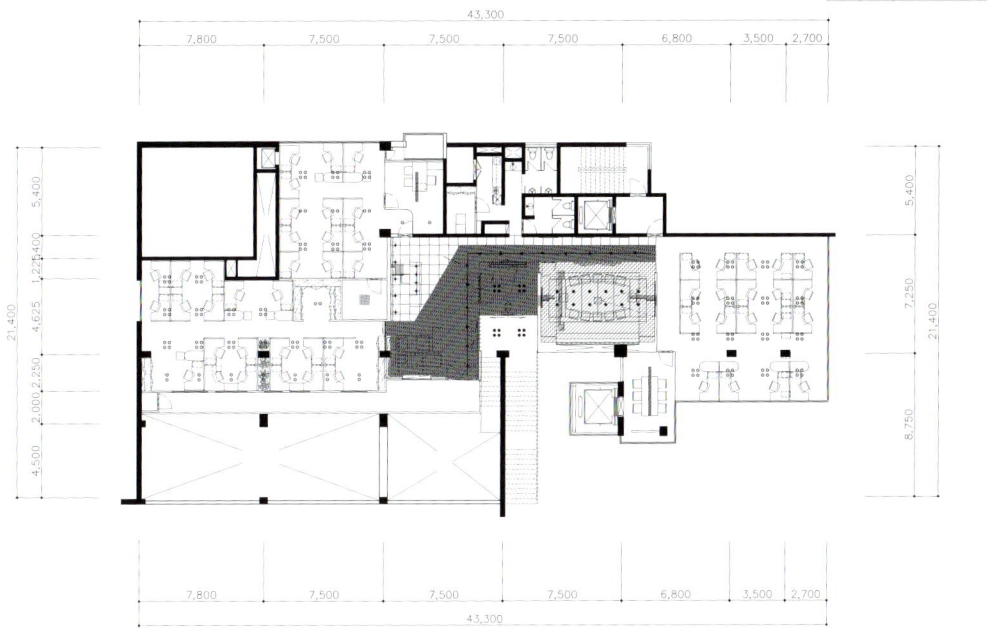

2층 천장도 / 2nd ceiling plan

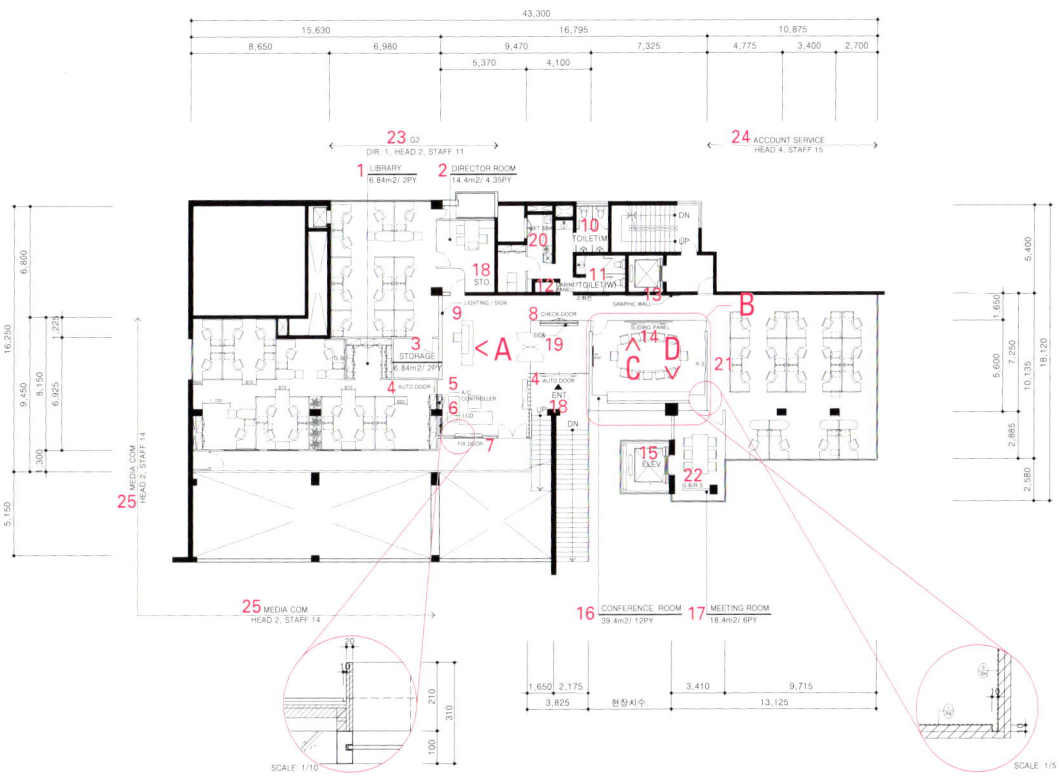

2층 평면도 / 2nd floor plan

1 LIBRARY 2 DIRECTOR ROOM 3 STORAGE 4 AUTO DOOR 5 A/C CONTROLLER 6 LCD 7 FIX. DOOR 8 CHECK DOOR 9 LIGHTING / SIGN 10 TOILET(M) 11 TOILET(W) 12 CABINET PANEL 13 GRAPHIC WALL 14 SLIDING PANEL 15 ELEV 16 CONFERENCE ROOM 17 MEETING ROOM 18 ENT 19 SIGN 20 WET SINK 21 R.S 22 G.B/R.B 23 G2 24 ACCOUNT SERVICE 25 MEDIA COM

1 서고 2 이사실 3 창고 4 자동문 5 에어컨 제어기 6 LCD 7 고정 문 8 점검 문 9 조명 / 사인 10 남자 화장실 11 여자 화장실 12 캐비닛 패널 13 그래픽 벽 14 미닫이 패널 15 엘리베이터 16 대회의실 17 소회의실 18 입구 19 사인 20 싱크 21 R.S 22 G.B/R.B 23 G2 24 회계팀 공간 25 미디어 컴

1 APP. COLOR PAINTED 2 APP. CORIAN FIN. 3 APP. S.ST'L H/L PLATE FIN. 4 DIA 50MM WARE HOLE FIN. 5 APP. CORIAN COLOR PAINTED 6 CABLE GUIDE : APP. CORIAN COLOR PAINTED 7 AUTO DOOR 8 APP. VINYL LEATHER FIN. 9 SEPARATION 10 T1.2 S.ST'L H/L PLATE FIN. 11 SLIDING PANEL 12 PDP 13 CLEAR GLASS 14 R.S

1 지정 컬러 도장 2 지정 코리안 마감 3 지정 스테인레스 스틸 헤어라인 판 마감 4 지름 50MM 철망 구멍 마감 5 지정 코리안 컬러 페인트 6 케이블 가이드 : 지정 코리안 컬러 페인트 7 자동문 8 지정 비닐 가죽 마감 9 분리대 10 T1.2 스테인레스 스틸 헤어라인 판 마감 11 미닫이 판넬 12 PDP 13 투명 유리 14 R.S

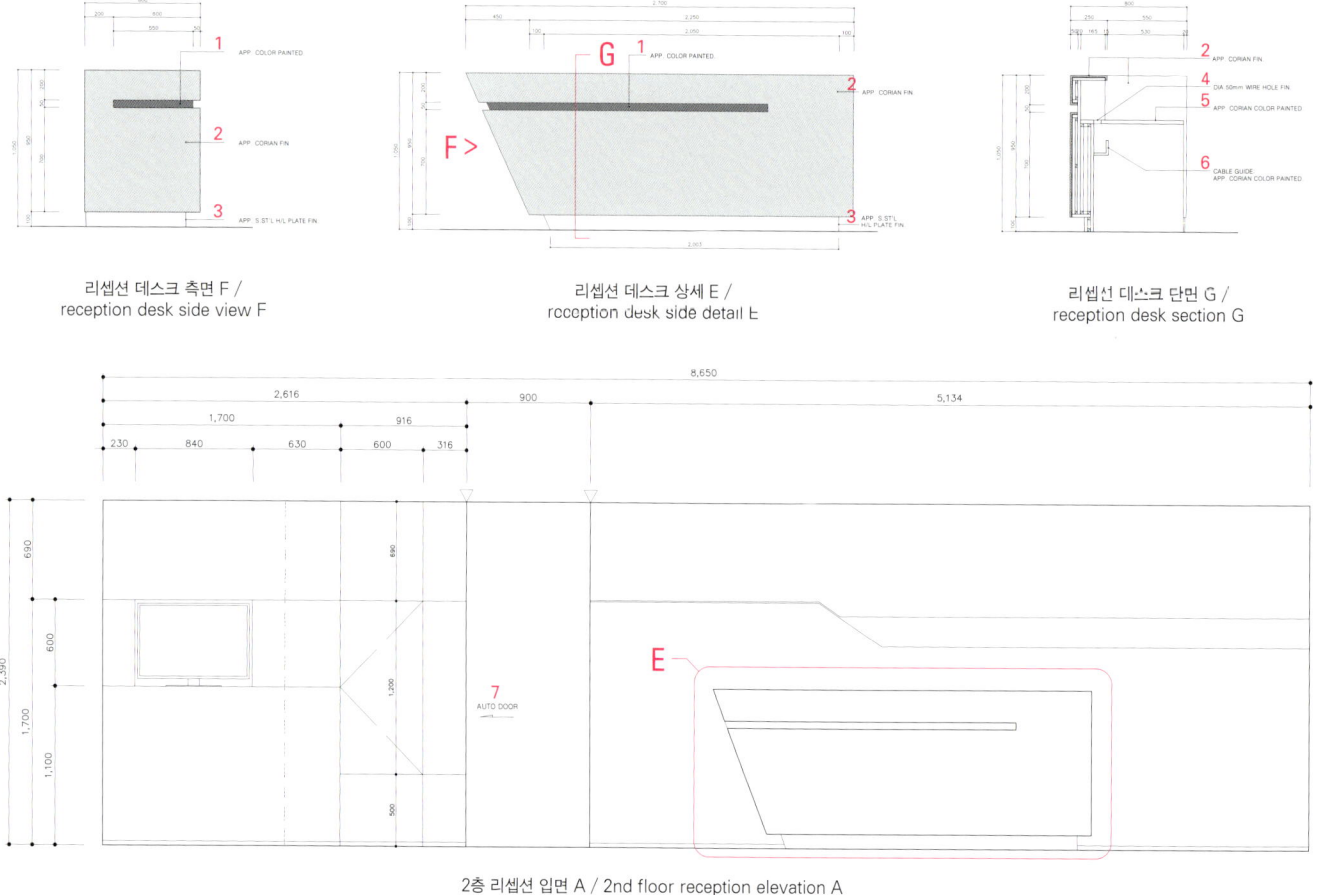

리셉션 데스크 측면 F / reception desk side view F

리셉션 데스크 상세 E / reception desk side detail E

리셉션 데스크 단면 G / reception desk section G

2층 리셉션 입면 A / 2nd floor reception elevation A

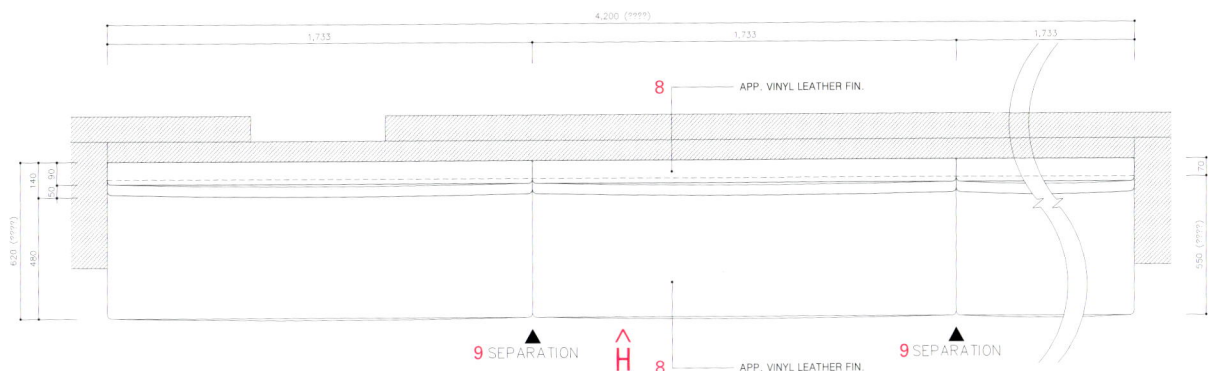

회의실 부스 소파 평면 / conference room booth sofa plan

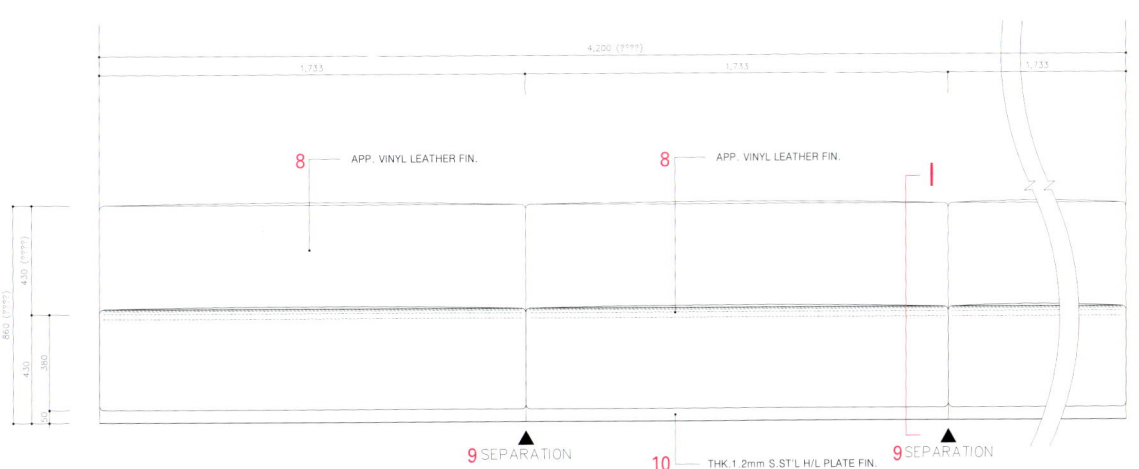

부스 소파 정면 H / booth sofa front view H

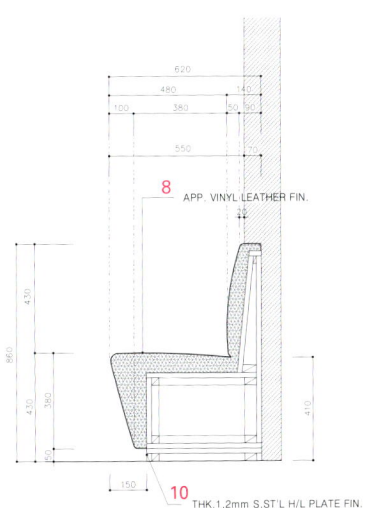

부스 소파 단면 I / booth sofa section I

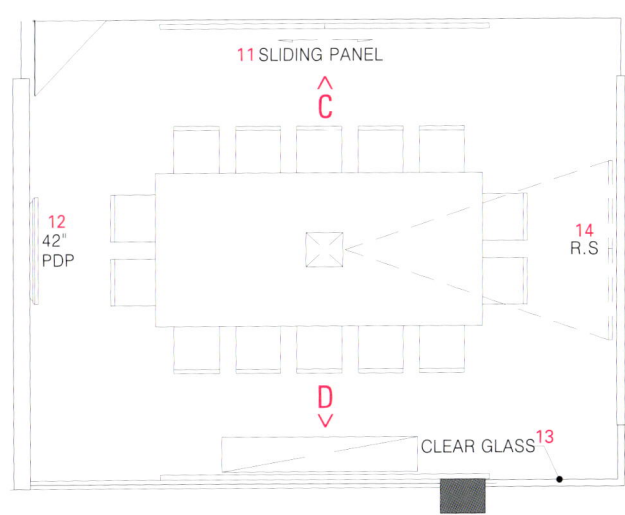

회의실 부분 평면 B / conference room partial plan B

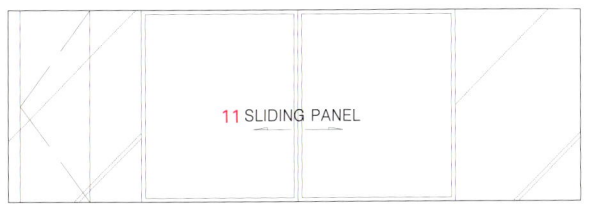

회의실 입면 C / conference room elevation C

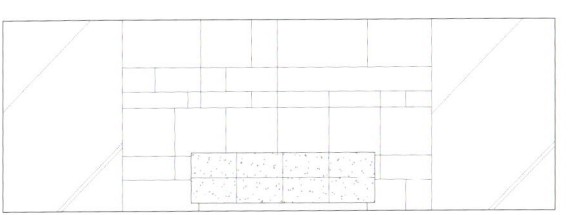

회의실 입면 D / conference room elevation D

1 DIRECTOR ROOM 2 CANTEEN 3 CONFERANCE ROOM 4 STORAGE 5 MOBILE RACK 6 BALCONY 7 O.A ROOM 8 IMAGE WALL 9 BOILER 10 TOILET(M) 11 TOILET(W) 12 ELEV 13 CEO ROOM 14 CABINET 15 WOOD BLIND 16 ROLL BLIND 17 PLANNING TEAM 18 MEETING ROOM 19 FIREPROOF SHUTTER 20 G.B 21 BOOK SHELF 22 BENCH 23 PUSH BALL 24 METAL BLIND 25 OPEN CABINET 26 PARTITION 27 CREATIVE TEAM 28 FINANCE TEAM 29 APP. BRICH PLYWOOD W/CLEAR LACQ. FIN. 30 SEPARATION 31 V-CUTTING

1 이사실 2 탕비실 3 대회의실 4 창고 5 이동식 선반 6 발코니 7 O.A 실 8 이미지 월 9 보일러 10 남자 화장실 11 여자 화장실 12 엘리베이터 13 CEO실 14 장 15 우드 블라인드 16 롤 블라인트 17 기획팀 18 소회의실 19 방화셔터 20 G.B 21 책장 22 벤치 23 푸시볼 24 금속 블라인드 25 열린 캐비닛 26 파티션 27 크리에이티브 팀 28 재무팀 29 지정 자작나무 합판 W/ 투명 락카 마감 30 분리대 31 V-커팅

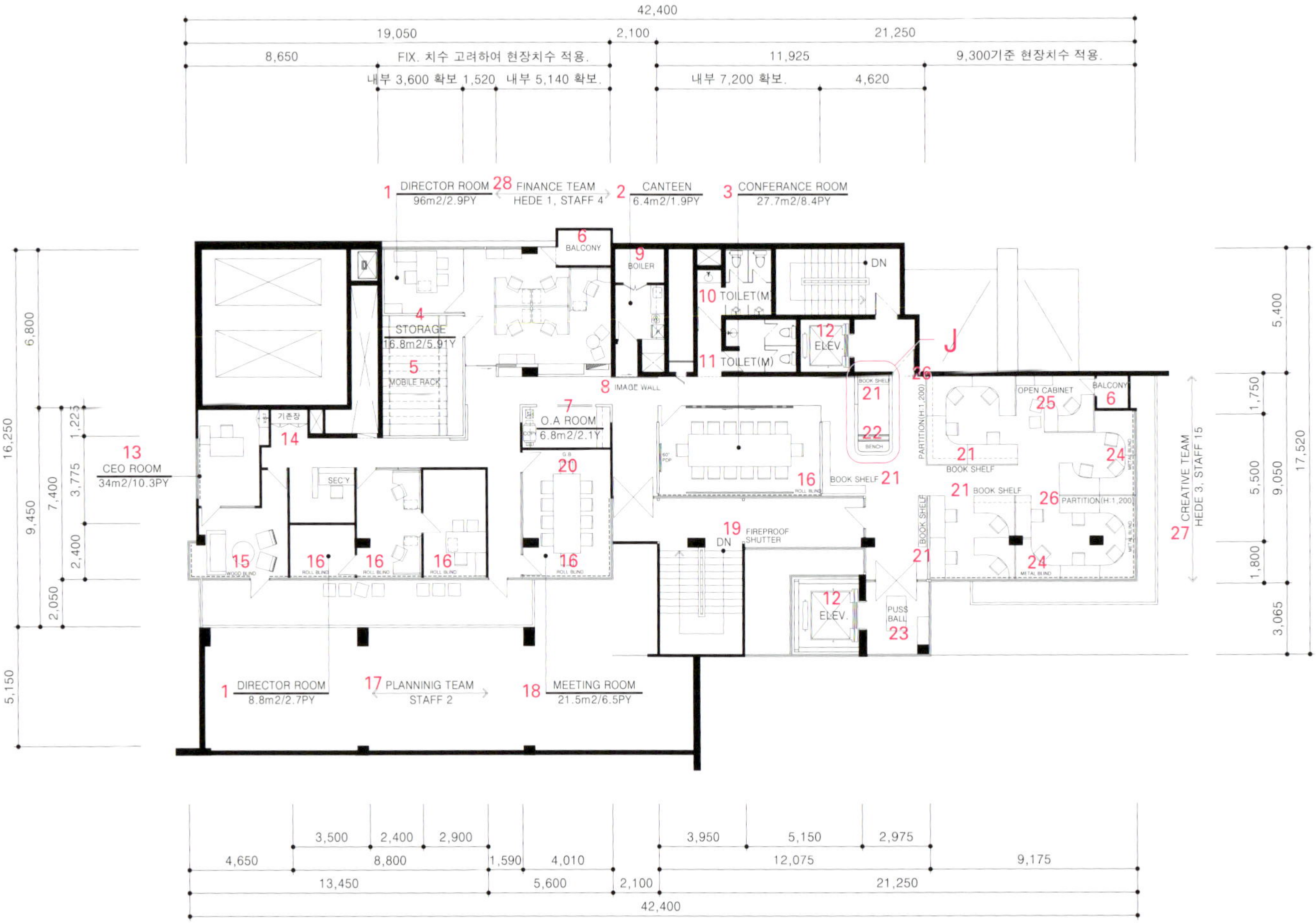

3층 평면도 / 3rd floor plan

3층 크리에이티브 팀 벤치 평면 J / 3rd floor creative team bench plan J

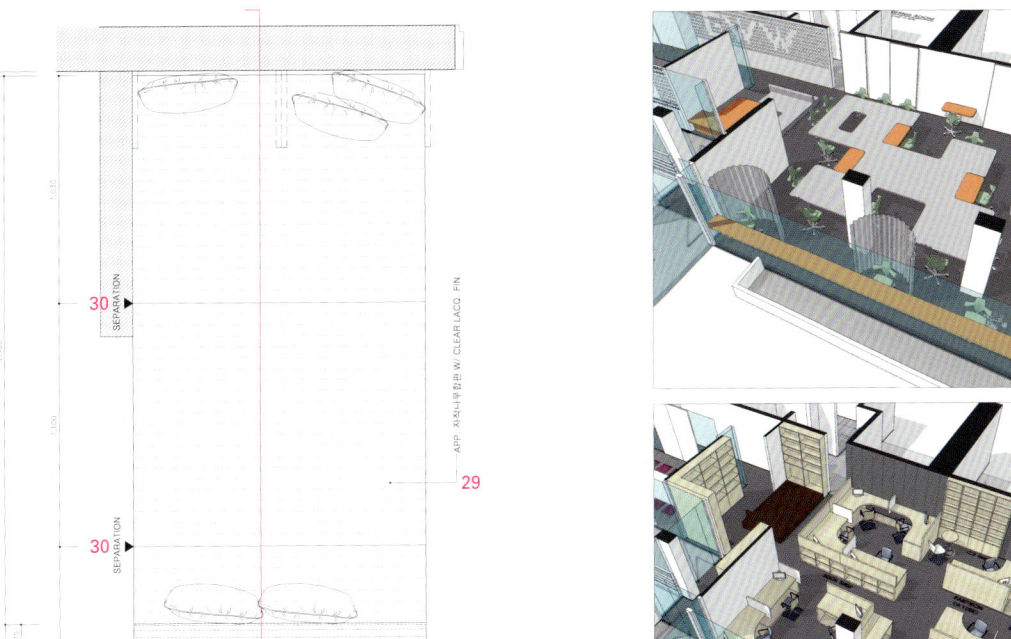

3층 크리에이티브 팀 엑소노메트릭 /
3rd floor creative team axonometric

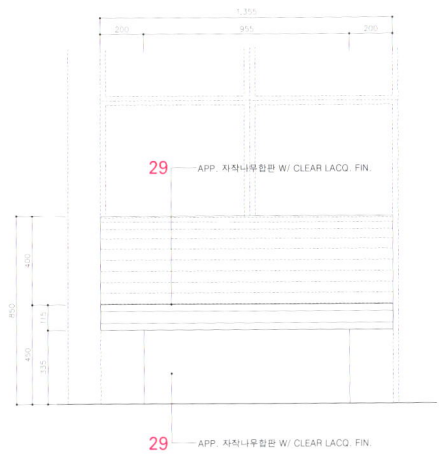

벤치 정면 K / bench front view K

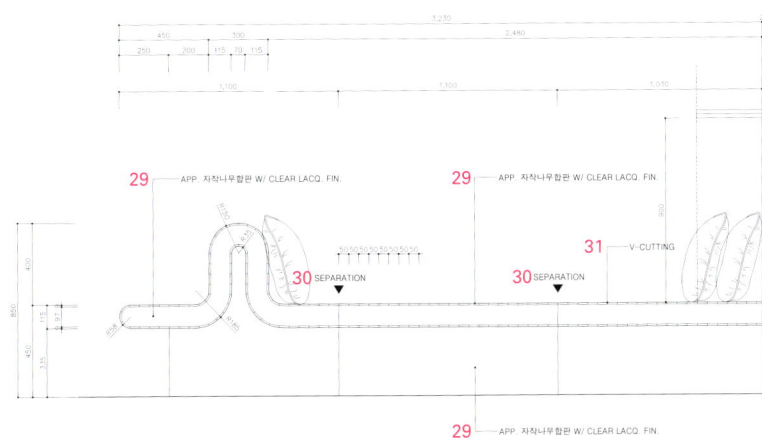

벤치 단면 L / bench section L

LS NETWORKS OFFICE

(주)다원디자인 | 조서윤 Dawon Design Co., Ltd. | Fay Cho

"변화와 혁신을 통해 비전을 현실로, 삶의 가치를 높여나가는 LS 네트웍스"라는 LS 네트웍스의 비전과 걸맞은 'New Wave / Wind of change'로 키워드를 잡고 컨셉을 풀어나갔다. LS 네트웍스는 크게 리셉션 / 디자이너 작업실 / 샘플실 / 라운지 / 열린 사무실 / 관리공간으로 구분하였다.
리셉션 공간은 사무실의 얼굴이 되는 부분으로 그 회사의 성격을 가장 잘 나타낼 수 있는 부분이다. 또한 라운지 개념으로 사람들과의 소통, 머묾의 공간이다. 브랜드 이미지를 극대화하기 위하여 다이나믹한 곡선을 이용하여 매스를 완성 시키고 이러한 요소들을 부각시키기 위해 오픈천장과 C.I 컬러를 적극적으로 활용하였다. 또한 이 공간을 마무리하는 측면에서 가구들의 배치나 형태를 기존의 획일화된 디자인의 가구에서 탈피하여 맞춤 주문의 의자와 외국 유명디자이너의 가구들을 직접 수입하여 적용하였다.
마지막으로 개방감과 확장성을 느낄 수 있도록 오픈천정으로 마감하고 천정의 Red arrow를 통해 개척자의 정신을 형상화하였다.

To accord with the vision of LS Networks "LS Networks, where vision becomes reality through change and innovation and life values are raised", 'new wave/wind of change' became the key words in finding the concept. LS Networks is mainly divided into the reception area, designer workroom, sample room, lounge, open office, and management space. The reception area is the face of the office and represents the character of the company in the best possible way. It also holds the concept of 'lounge', where people communicate and spend some time. To maximize the brand image, dynamic, curved lines completed the mass and to stress these factors, an open-ceiling and C.I. colors were used aggressively. To finish this interior space and avoid common furniture and arrangement practices, custom-made chairs and furniture designed by famous foreign designers were positioned.

Finally, there is an open ceiling that creates a feeling of openness and expansiveness, further heightened by the red arrow on the ceiling that represents pioneering spirit.

위치 서울시 용산구 한강로2가 LS용산타워 11, 12층
면적 800평
설계 및 시공 (주)다원디자인 / 조서윤, 남용식, 김재환
마감 리셉션 : 바닥 – 카페트 타일, 세라믹 타일, 벽 – 컬러 페인트, 컬러 시트, 천정 – 노출 천장, 컬러 페인트(적색), 스트라이프 조명 박스
사진 (주)다원디자인 제공

Location 11,12th LS Yongsan Tower, Hangangno 2-ga, Yongsan-gu, Seoul, Korea
Area 800py
Construction & Design Dawon Design Co., Ltd. / Fay Cho, Nam Yong Sik, Kim Jae Hwan
Finishing Reception : Floor - Carpet Tile , Ceramic Tile, Wall - Color Paint , Color Sheet, Ceiling - Open ceiling, Color paint(Red), Stripe Lighting Box
Photos offer Dawon Design Co., Ltd.

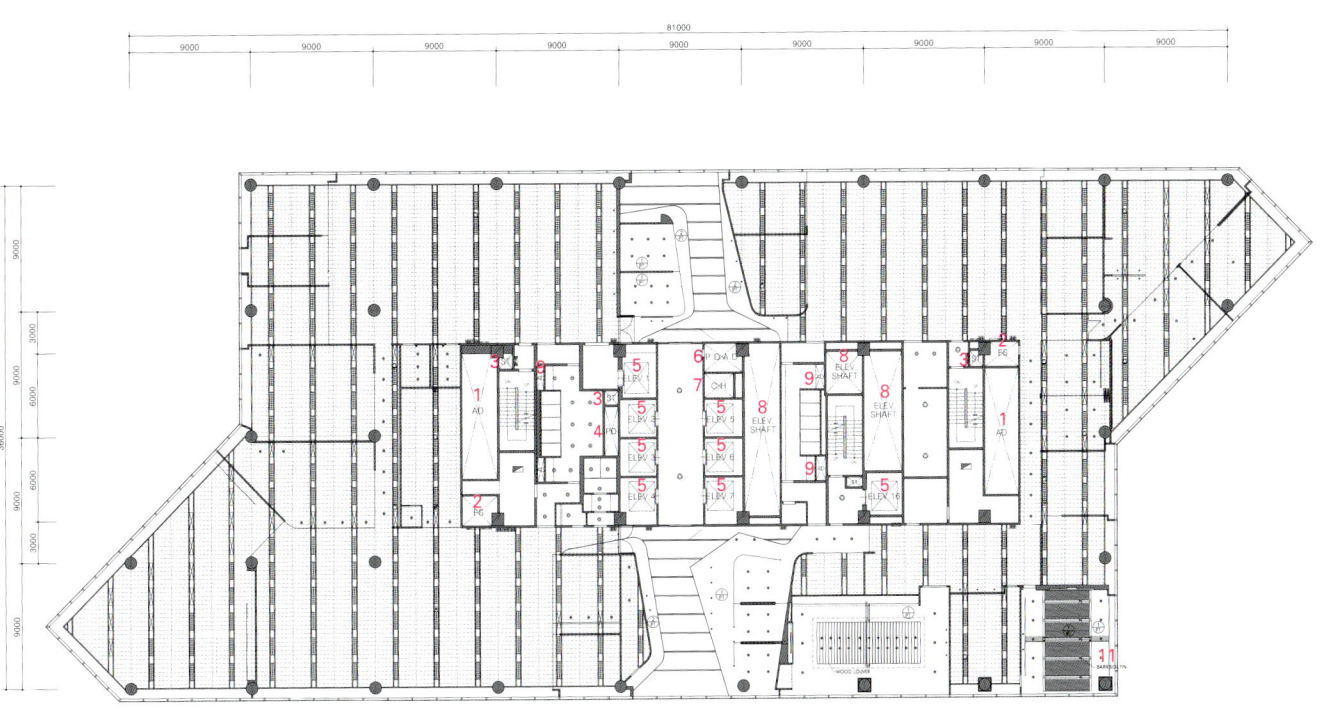

12층 천장도 / 12th ceiling plan

12층 평면도 / 12th floor plan

1 AD 2 FS 3 ST 4 PD 5 ELEVATOR 6 PD / AD 7 CH 8 ELEV. SHAFT 9 GRAPHIC 10 방화문 11 BARRISOL FIN. 12 WOOD LOUVER 13 SIGN BOARD 14 AUTO DOOR

1 AD 2 FS 3 ST 4 PD 5 엘리베이터 6 PD / AD 7 CH 8 엘리베이터 통로 9 그래픽 10 방화문 11 바리솔 마감 12 목재 루버 13 사인보드 14 자동문

1 INDRECTED LIGHTING 2 RED COLOR 3 DOOR TO OPEN OFFICE 4 OPEN TO CORRIDOR 5 T1 FRAME 6 T10 BASE 7 OPEN TO MTG RM 5 8 OPEN TO MTG RM 6 9 APP. WOOD VENEER FIN. 10 DIA. 50MM WIRE HOLE W/ S.ST'L COVER FIN. 11 APP. CORIAN FIN. 12 CABLE GUIDE : APP. WOOD VENEER FIN. 13 APP. S.ST'L POLISHED PLATE FIN. 14 MOVABLE DRAWER BOX : 420X500X680 15 APP. INDIRECT LIGHTING FIN.

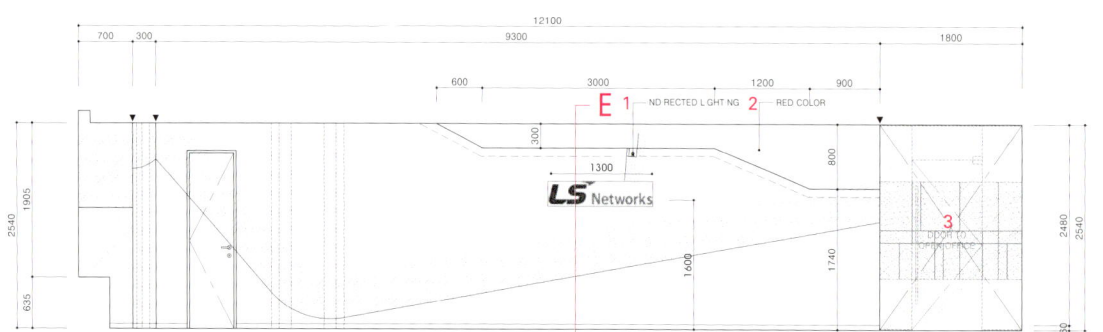

12층 리셉션 입면 A / 12th floor reception area elevation A

벽 단면 E / wall section E

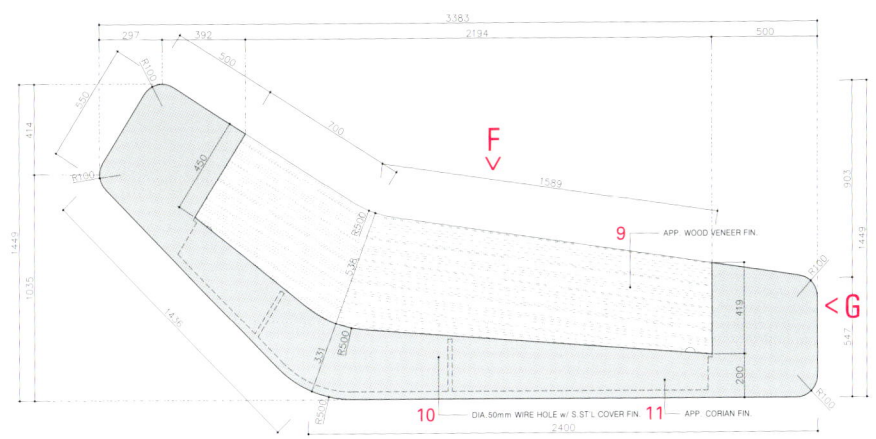

리셉션 데스크 평면 C / reception desk plan C

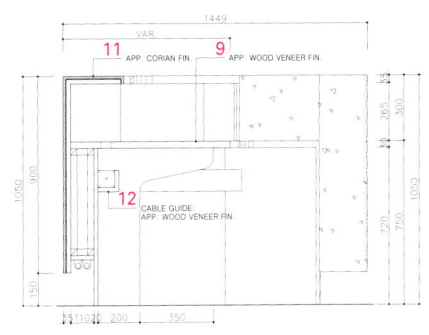

리셉션 데스크 단면 H / reception desk section H

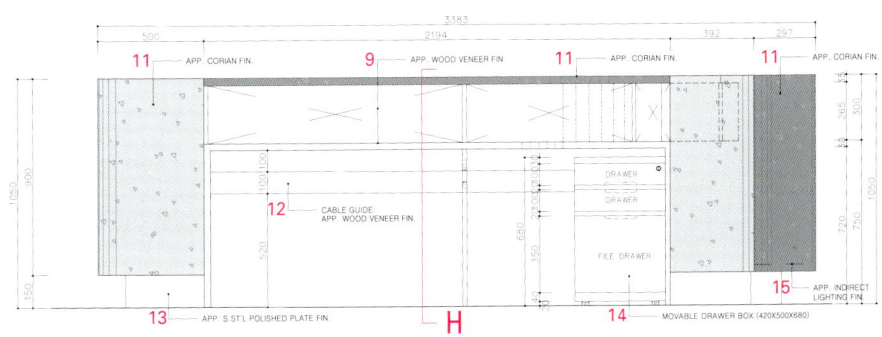

리셉션 데스크 후면 F / reception desk rear view elevation F

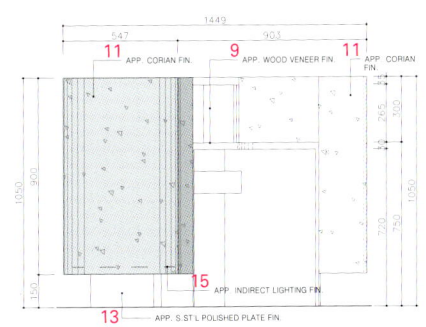

리셉션 데스크 측면 G / reception desk side veiw G

1 간접 조명 2 적색 3 사무실 쪽 문 4 복도 열린공간 5 T1 틀 6 T10 베이스 7 회의실 5 열린공간 8 회의실 6 열린공간 9 지정 무늬목 마감 10 지름 50MM 철선 구멍 W/ 스테인레스 스틸 피복 마감 11 지정 코리안 마감 12 케이블 유도장치 : 지정 무늬목 마감 13 지정 스테인레스 스틸 광택 판 마감 14 이동식 서랍 : 420X500X680 15 지정 간접 조명 마감

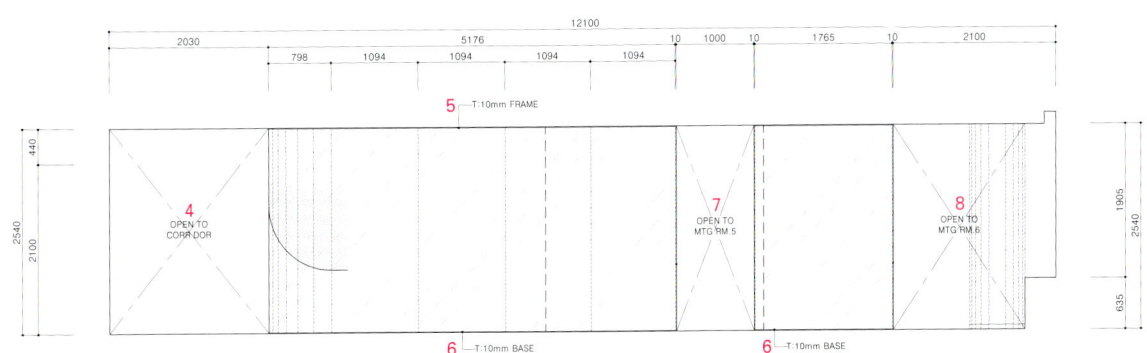

12층 리셉션 입면 B / 12th floor reception area elevation B

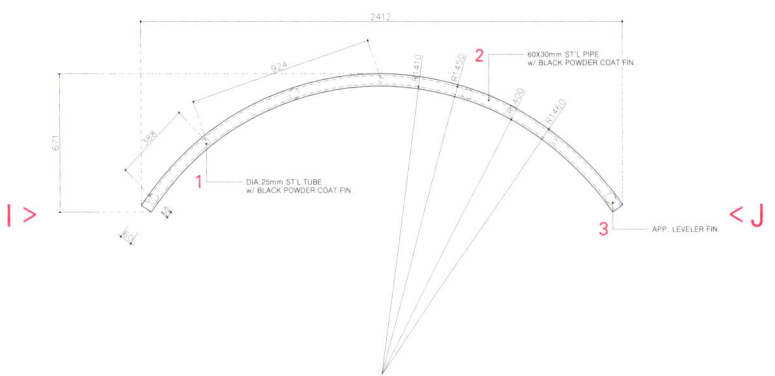

라운지 파티션 부분 평면 / lounge partition partial plan

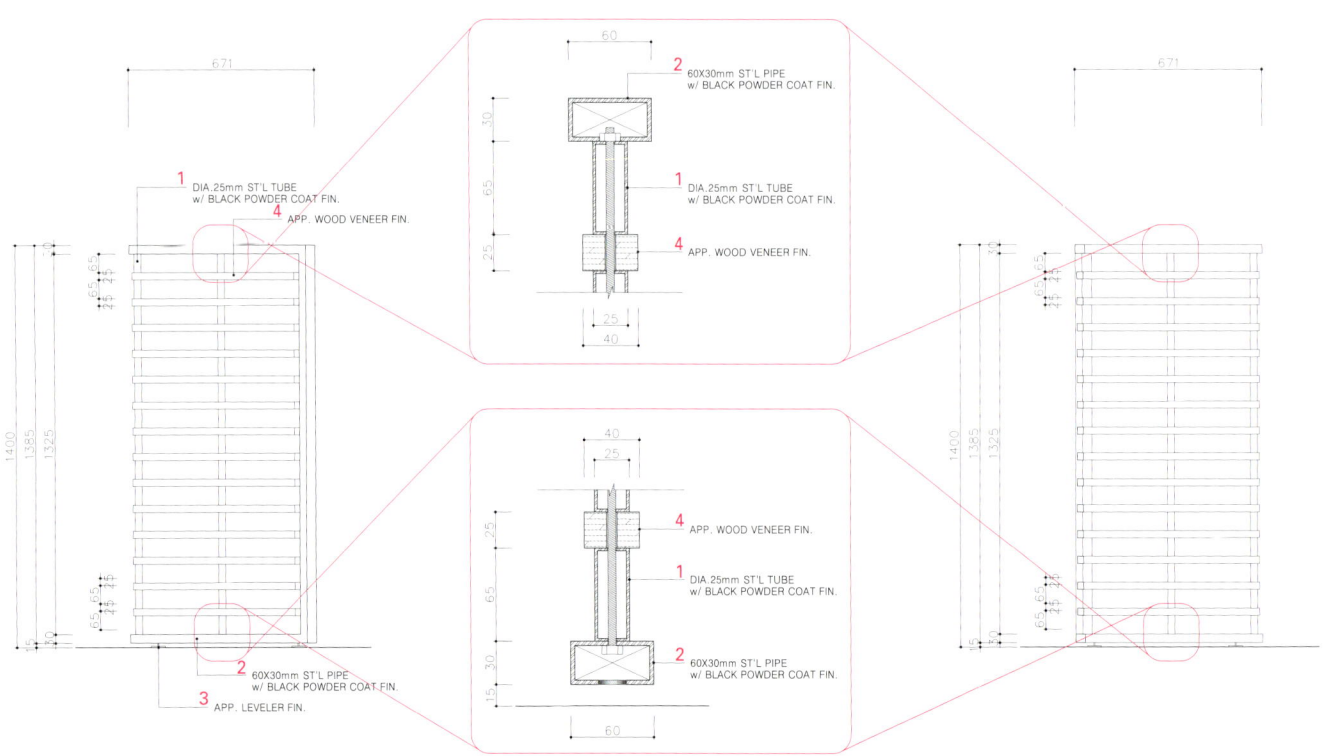

파티션 입면 I / partition elevation I

파티션 입면 J / partition elevation J

1 DIA. 25MM ST'L TUBE W/ BLACK POWER COAT FIN. 2 60X30 ST'L PIPE W/ BLACK POWDER COAT FIN. 3 APP. LEVELER FIN. 4 APP. WOOD VENEER FIN. 5 DECORATION SLOT 6 APP. LEATHER FIN. 7 400X400 CUSHION : APP. FABRIC FIN. 8 BASE LINE

1 지름 25MM 스틸 튜브 W/ 블랙 파우더 코팅 마감 2 60X30 스틸 파이프W/ 블랙 파우더 코팅 마감 3 지정 측량 마감 4 지정 무늬목 마감 5 데코레이션 슬롯 6 지정 가죽 마감 7 400X400 쿠션 : 지정 패브릭 마감 8 기준선

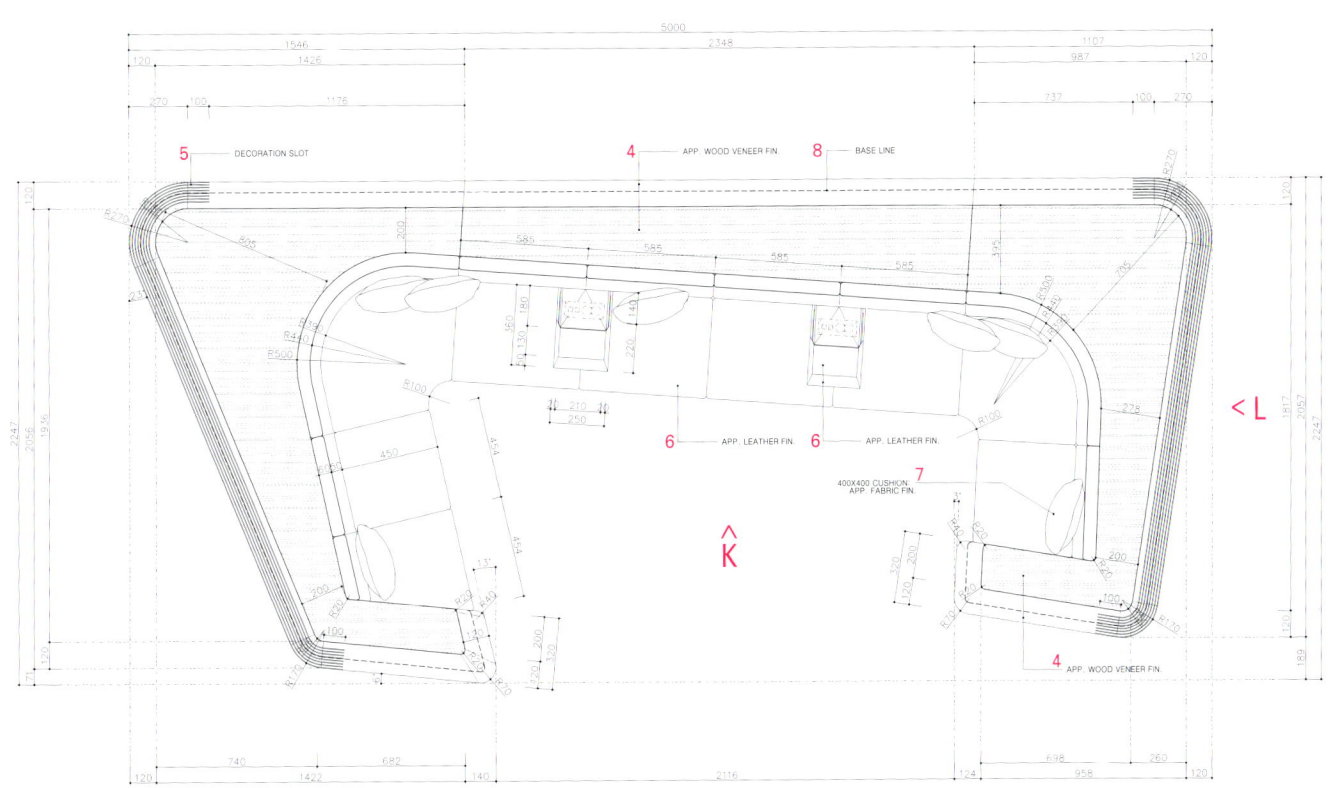

부스 소파 평면 D / booth sofa plan D

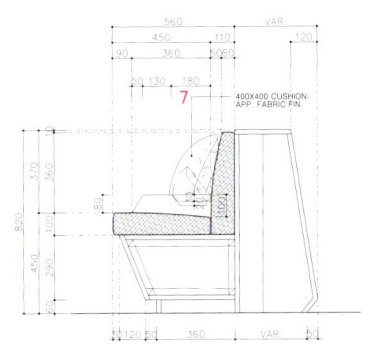

부스 소파 단면 M / booth sofa section M

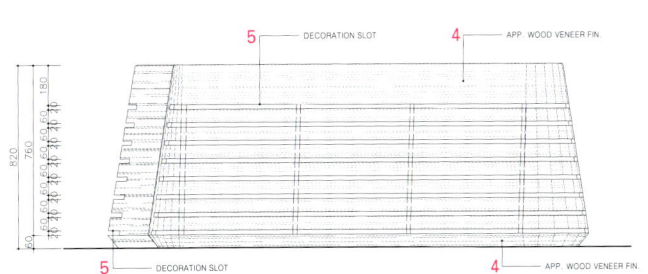

부스 소파 측면 L / booth sofa side view L

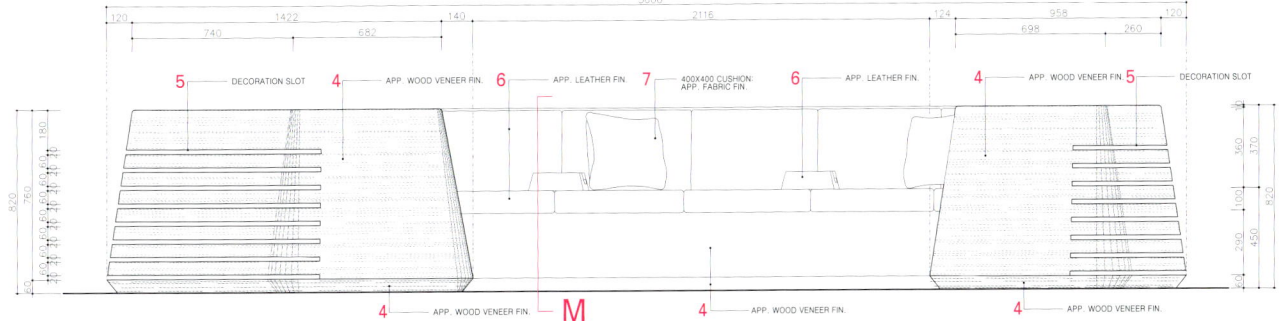

부스 소파 정면 K / booth sofa front view K

IBK PRIVATE BANKING CENTER

(주)다원디자인 | 조서윤　DAWON INTERIOR DESIGN | Fay Cho

IBK 기업은행의 Private Banking(PB)센터 1호점인 '강남PB센터' 의 설계시 주안점은 지역적 특색과 고객의 요구를 반영하여 다양한 문화와 만남의 공간을 제공하여 고객에게 즐거움을 줄 수 있는 공간으로 만드는 것이었다. 직원의 동선 및 효율성을 고려하면서 고객의 프라이버시를 위하여 상담실의 배치를 일반적이지 않은 창가 측에 복도를 두어 자연 채광과 외부의 뷰를 확보하였으며 4개의 각 상담실의 컨셉은 간접 조명이 있는 단순화된 모던 스타일의 한실 온돌 타입 상담실, 녹색 계열의 컬러풀 하면서 편안한 책상 배치의 전통적인 형태의 상담실 프렌치 거실 & 서재 타입의 고풍스러우며 따뜻한 서재 또는 응접실 용도의 2개의 다목적 상담실로 구성 하였다.

의도적으로 계획되어진 조금은 지루하게 보여질 수 있는 긴 복도 면에는 여성스러움과 부드러움을 강조한 천정과 벽체의 청동 금속 펀칭 패턴, 그래픽 이미지 & 목재 격자창을 사용하여 단조로움을 피하였으며 각 실, 공간별로 소품 및 제작 가구에도 기능에 충실하면서 디자인 표현에 중점을 두어 계획하였다. 또한 고객 편의를 위하여 파우더 & 화장실을 신설하였으며 그 외의 직원들을 위한 업무 공간은 독립성을 유지하면서 서로 유기적일수 있도록 공간들을 배치하여 편안함과 기능적일 수 있도록 계획 하였다.

he essential point in the design for the 1st branch of IBK(Industrial Bank of Korea) PB(Private Banking) center, 'Gangnam PB Center' was to reflect the regional characteristics and customer needs and provide a place for various cultures and meeting to make it a space where customers can genuinely enjoy their stay.
In consideration of staff traffic line and efficiency as well as customers' privacy, a corridor was put next to the window of consultation rooms to secure natural lighting and external views, which isn't a common arrangement, and four consultation rooms are a modernized traditional Korean ondol room(a room with a floor heating system) with indirect lighting, a typical consultation room with comfortable desk arrangement in green colors and two multipurpose consultation rooms that can also function as warm libraries or living rooms in an antique french living room & library style.
For the corridor that was intentionally planed in a long shape, a ceiling that emphasizes feminity and gentleness, bronze metal punching patterns on the wall and graphic images & wooden lattice windows were designed to avoid monotonousness and every item and furniture was scrupulously designed or manufactured based on its function and location. Besides powder & rest-rooms were newly established for customer convenience and business spaces for staff were arranged to be separated from each other and still connected together organically to hold both convenience and efficiency.

위치 서울시 강남구 도곡동 467-7 아카데미 스위트 2층
용도 은행
면적 339㎡(102.7평)
설계기간 2009. 6 ~ 2009. 7
공사기간 2009. 8 ~ 2009. 9
설계담당 (주)다원디자인 / 조서윤, 김주상, 노경옥
공사담당 (주)다원디자인 / 가창순
마감 바닥 – 대리석, 카펫, 우드 플로링, 러그, 벽 – 대리석, 패브릭, 무늬목, 슈퍼 그래픽, 벽지, 컬러 락카, 천정 – 페인트
사진 (주)다원디자인 제공

Location 2F, Academy sweet, 467-7 Dogok-dong, Gangnam-gu, Seoul, Korea
Use Private Banking
Area 339㎡(102.7PY)
Design period 2009. 6 ~ 2009. 7
Construction period 2009. 8 ~ 2009. 9
Design team DAWON INTERIOR DESIGN / Fay Cho, Kim Ju Sang, No Kyung Ok
Construction DAWON INTERIOR DESIGN / Ga Chang Soon
Finishing Floor - Marble, Carpet, Wood flooring, Rug, Wall - Marble, Fabric, Wood veneer, Super graphic, Wall covering, Color lacq, Ceiling - Paint
Photos offer DAWON INTERIOR DESIGN

천장도 / ceiling plan

평면도 / floor plan

1 GYPSUM 1PLY + MAITON 2 RUST-PREVENTING PANEL ON G.B WITH ENAMEL PAINT 3 INDIRECT LIGHT'G 4 GALVA ON COATING FIN. 5 CEILING FRAME RAIL 6 FIREPROOF SHUTTER 7 10MM REVEAL 8 WOOD SHEET 9 RUST-PREVENTING PANEL ON G.B WITH WALLPAPER 10 AIR-SUPPLY & VENTILATION LOUVER(CONSTRUCTION AFTER REMOVING EXIST WINDOW) 11 AIR-SUPPLY & VENTILATION LOUVER 12 ARTWORK 13 PBA 14 TEA ROOM 15 BLIND 16 SPECIALIST CONSULTING ROOM 17 CREDENZA 18 ARTWORK 19 LOCKER 20 TECHNICAL ASSISTANCE ROOM 21 VAULT 22 OA 23 SERVICE ROOM 24 MOVABLE PARTITION 25 INFO 26 POLICE 27 FILE CABINET 28 P.B RM 29 CENTER CHIEF ROOM 30 WOOD BLIND 31 LIVING & LIBRARY 32 TOILET(W) 33 AUTO DOOR 34 FIRE EXTINGUISHER 35 WATERPROOFING FIXTURES BOX 36 FRONT ROOM 37 LOAN CASHBOX 38 UPSROOM 39 POWDER ROOM

1 석고 1겹 + 마이톤 2 내화판넬 위 석고보드 위 에나멜 페인트 3 간접 조명 4 갈바 위 도장마감 5 천정 액자 레일 6 방화 셔터 7 10MM 창틀 8 목재 시트 9 내화판넬 위 석고보드 위 벽지 10 급배기용 루바(기존 창호 철거 후 루바 시공) 11 급배기용 루바 12 공예품 13 PBA 14 다실 15 블라인드 16 전문가 상담실 17 식기장 18 공예품 19 락카 20 전산장비실 21 금고실 22 OA 23 탕비실 24 이동식 파티션 25 안내 26 청경 27 화일 캐비닛 28 P.B 실 29 센터장실 30 우드 블라인드 31 거실 & 서재 32 여자 화장실 33 자동문 34 소화전 35 방수 기구함 36 전실 37 대여금고 38 UPS실 39 파우더 룸

IBK PRIVATE BANKING CENTER | 기업은행 강남 PB 센터

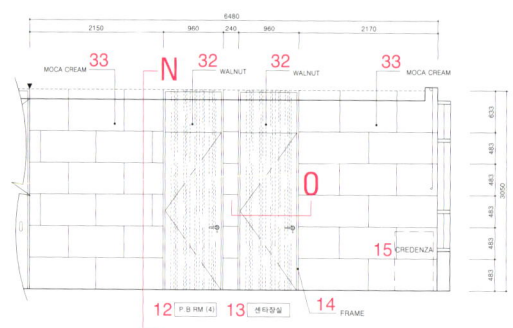

복도 입면 B / corridor elevation B

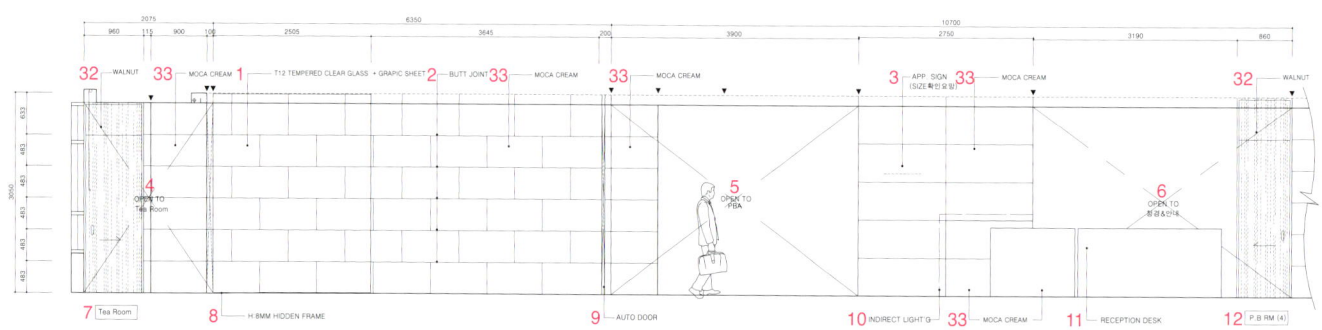

리셉션 복도 입면 A / reception corridor elevation A

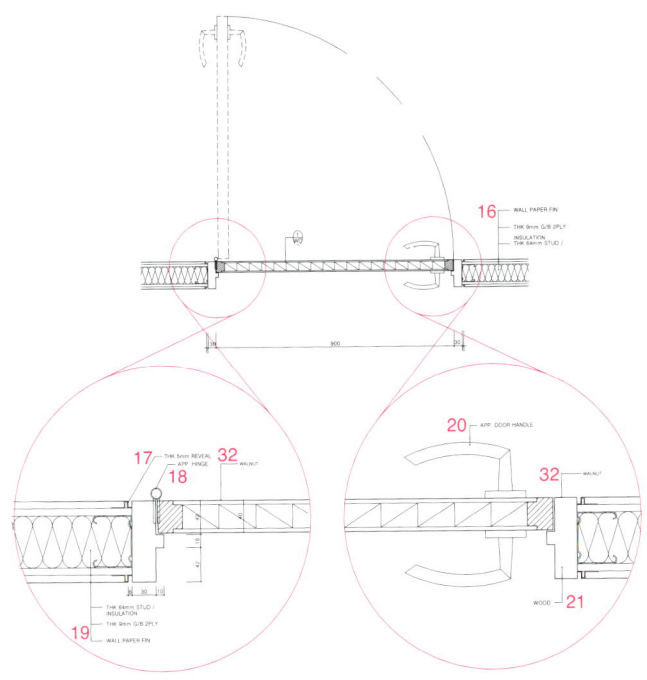

문 단면 상세 O / door section detail O

1 T12 TEMPERED CLEAR GLASS + GRAPIC SHEET 2 BUTT JOINT 3 APP. SIGN 4 OPEN TO TEA ROOM 5 OPEN TO PBA 6 OPEN TO POLICE & INFO 7 TEA ROOM 8 H=8MM HIDDEN FRAME 9 AUTO DOOR 10 INDIRECT LIGHT'G 11 RECEPTION DESK 12 P.B RM 13 CENTER CHIEF ROOM 14 FRAME 15 CREDENZA 16 WALL PAPER FIN. / T9 G/B 2PLY / INSULATION T64 STUD 17 T5 REVEAL 18 APP. HINGE 19 T64 STUD, INSULATION / T9 G/B 2PLY / WALL PAPER FIN. 20 APP. DOOR HANDLE 21 WOOD 22 PBA 23 DECO LAMP 24 T9.5 G.B 2PLY V.P FIN. 25 CORRIDOR 26 BUTT JOINT 27 ㅁ-75X45 ST PIPE / T9 PLYWOOD / 5MM EPOXY BOND / COPPER WIRE REINFORCING / T20 STONE FIN. 28 T10 RUG CAPET / T15 FELT 29 5X20 SST'L MOULDING 30 T1.6 GALVA ST'L, BRONZE COLOR PAINT / T12 TEMPERED CLEAR GLASS 31 30X45 BRONZE COLOR SUS FRAME 32 WALNUT 33 MOCA CREAM

1 T12 투명 강화 유리 + 그래픽 시트 2 맞대이음 3 지정 사인 4 다실 열린공간 5 PBA 열린공간 6 청경 & 안내 열린공간 7 다실 8 H=8MM 숨은 틀 9 자동문 10 간접 조명 11 리셉션 데스크 12 P.B 실 13 센터장실 14 틀 15 식기장 16 벽지 마감 / T9 석고보드 2겹 / 단열재 T64 스터드 17 T5 창틀 18 지정 힌지 19 T64 스터드, 단열재 / T9 석고보드 2겹 / 벽지 마감 20 지정 문 손잡이 21 목재 22 PBA 23 데코 램프 24 T9.5 석고보드 2겹, 도장마감 25 복도 26 맞대이음 27 ㅁ-75X45 스틸 파이프 / T9 합판 / 5MM 에폭시 본드 / 동선 보강 / T20 석재마감 28 T10 러그카펫/ T15 펠트 29 5X20 스테인레스 스틸 몰딩 30 T1.6 갈바 스틸, 청동 컬러 페인트 / T12 투명 강화 유리 31 30X45 청동 컬러 스테인레스 프레임 32 월넛 33 모카크림색

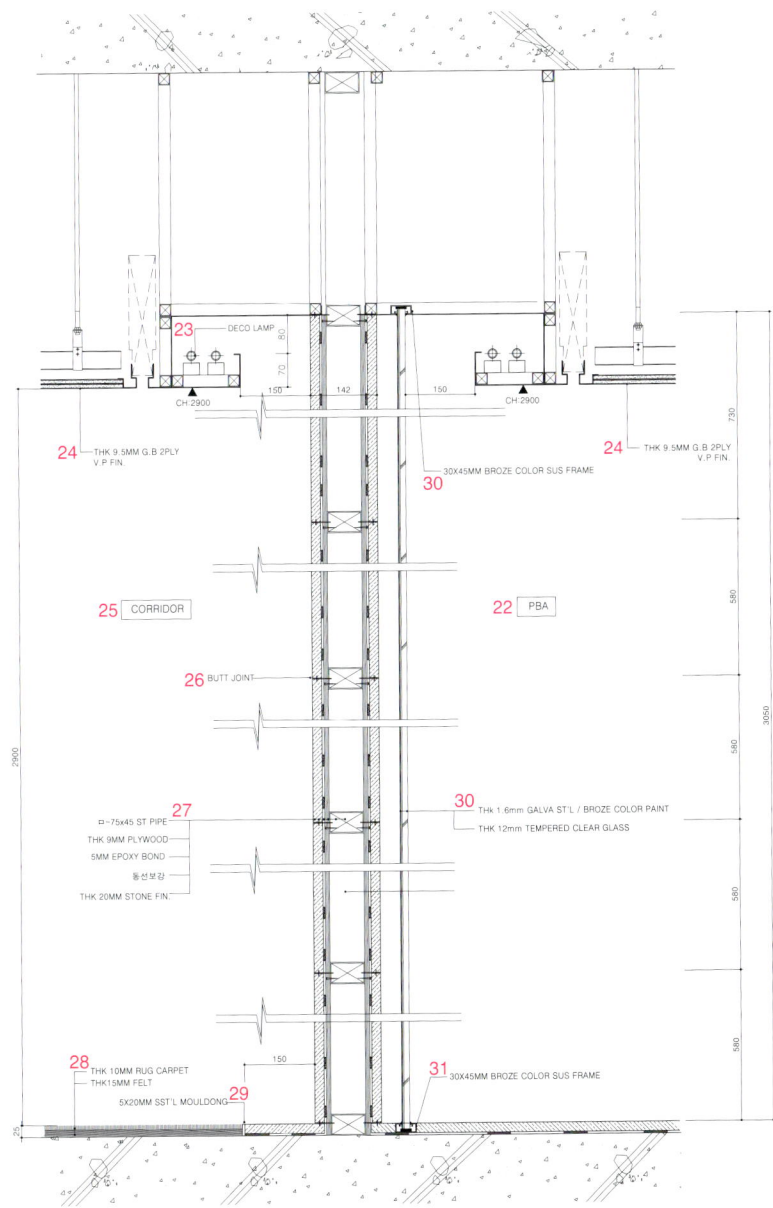

벽 단면 N / wall section N

1 APP. STONE FIN. 2 APP. WOOD VENEER FIN. 3 APP. WOOD VENEER W/ T5 TEMPERED CLEAR GLASS FIN. 4 PC BODY 5 APP. INDIRECT LIGHTING FIN. 6 LAMINATED GLASS 7 APP. S.ST'L H/L FIN. 8 APP. INDIRECT LIGHTING FIN. 9 PRINTER 10 OPEN 11 CREDENZA 12 AL. LOUVER / BENJAMIN MOORE NAVAJO WHITE 73 13 POWDER ROOM / TOILET(W) 14 FRAME 15 DP 16 FRONT ROOM 17 AUTO SLIDING DOOR CHECK POINT 18 STEEL / WOOD SHEET 19 AUTO SLIDING DOOR 20 T1.6 GALVANIZED PLATE V.P FIN. 21 T9.5 G.B 2PLY V.P FIN. 22 OPEN TO LIVING & LIBRARY 23 STEEL / WOOD FILM 24 AUTO DOOR 25 WALNUT 26 BENJAMIN MOORE CHINA WHITE 74

1 지정 석재 마감 2 지정 무늬목 마감 3 지정 무늬목 마감/ T5 투명 강화 유리 마감 4 PC 본체 5 지정 간접 조명 마감 6 접합유리 7 지정 스테인레스 스틸 헤어라인 마감 8 지정 간접 조명 마감 9 프린터 10 열림 11 식기장 12 알루미늄 루버 / 백색 도장 (벤자민 무어 노바조 화이트 73) 13 파우더룸 / 여자화장실 14 틀 15 DP 16 전실 17 자동 미닫이 문 점검구 18 스틸 / 목재 시트 19 자동 미닫이 문 20 T1.6 아연도금 판 도장 마감 21 T9.5 석고보드 2겹, 도장 마감 22 거실 & 서재 열린공간 23 스틸 / 목재 필름 24 자동문 25 월넛 26 백색 도장(벤자민 무어 차이나 화이트 74)

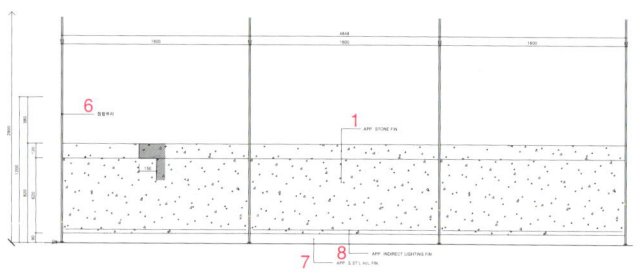

PBA 카운터 정면 E / PBA counter front view E

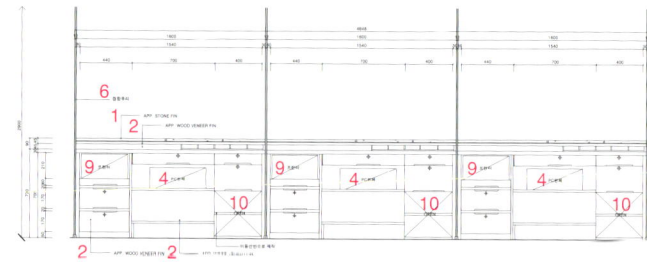

PBA 카운터 후면 E / PBA counter rear view E

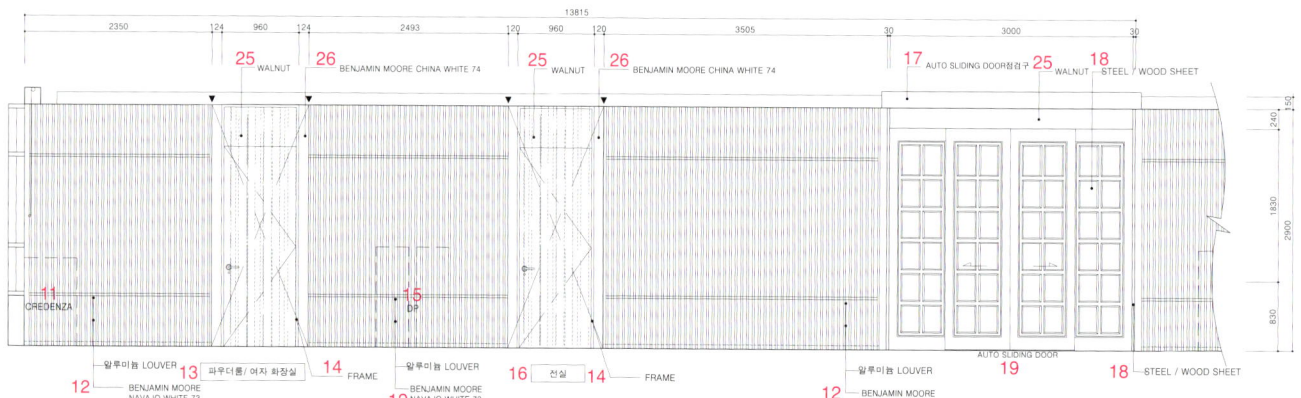

복도 입면 C / corridor elevation C

1 CALKING 2 T65 STUD, T50 INSULATION / T9.5 GYPSUM BOARD 2PLY 3 T9.5 G.B 2PLY V.P FIN. 4 FRONT ROOM 5 T20 MARBLE FIN. / T5 ADHESIVE 6 WOOD FLOORING 7 WOOD VENEER 8 T9.5 G.B 2PLY V.P FIN. 9 TEA ROOM 10 T10X10 WOOD 11 T3 KOREAN PAPER ACRYL / WOOD DOOR FRAME 12 WOOD FLOORING / T15 ELECTRIC ONDOL SYSTEM(READY-MADE) / T12 WATERPROOFING PLYWOOD / T3 DUSTPROOF RUBBER / T12 WATERPROOFING PLYWOOD / T3 DUSTPROOF RUBBER / 30X30 SQUARED PIPE 13 SLIDING DOOR 14 T9 REVEAL 15 INDIRECTED LIGHT'G 16 T6 REVEAL / T50 BASE 17 ARTWORK 18 CREDENZA 19 9MM GYPSUM BOARD 2PLY / 6MM PLYWOOD, WALLPAPER FIN. 20 TEA RM(KOREAN PAPER ACRYL) 21 102X32 STUD, 12MM GYPSUM BOARD 2PLY 22 FRAME 23 WOOD 24 PINE NEEDLE PERFUME KOREAN WALLPAPER #2007-1 25 WALNUT 26 T5 MIRROR

1 코킹 2 T65 스터드, T50 단열재 / T9.5 석고보드 2겹 3 T9.5 석고보드 2겹, 도장 마감 4 전실 5 T20 대리석 마감 / T5 접착제 6 우드 플로링 7 무늬목 8 T9.5 석고보드 2겹, 도장 마감 9 다실 10 T10X10 원목 11 T3 한지 아크릴 / 목재 문틀 12 우드 플로링/ T15 전기온돌시스템(기성재) / T12 내수합판 / T3 방진고무 / T12 내수합판 / T3 방진고무 / 30X30 각파이프 13 미닫이문 14 T9 창틀 15 간접 조명 16 T6 창틀 / T50 베이스 17 공예품 18 식기장 19 9MM 석고보드 2겹 / 6MM 합판, 벽지마감 20 다실(한지 아크릴) 21 102X32 스터드, 12MM 석고보드 2겹 22 프레임 23 원목 24 솔잎향 한지벽지 #2007-1 25 월넛 26 T5 거울

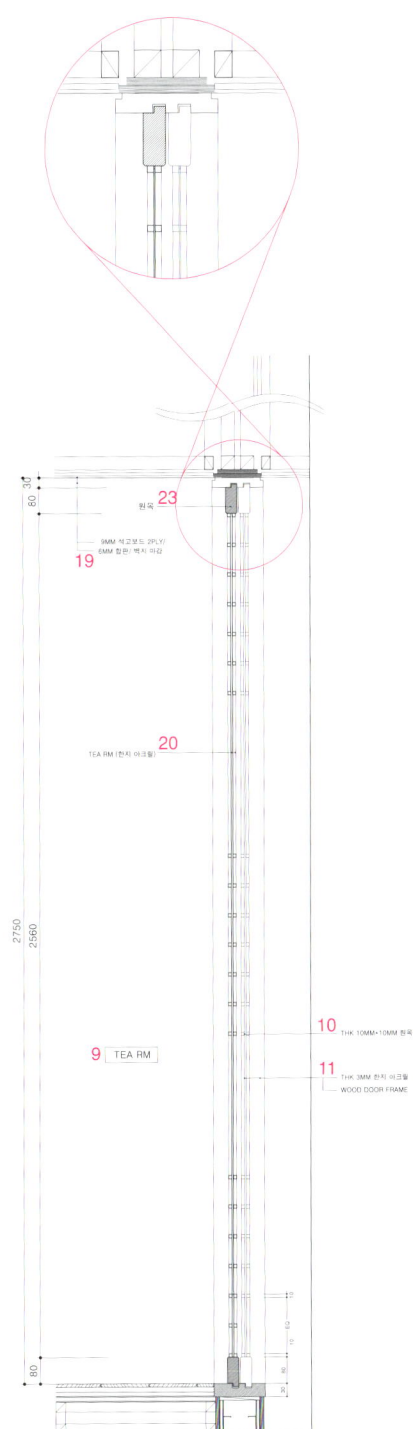

문 단면 상세 U / door section detail U

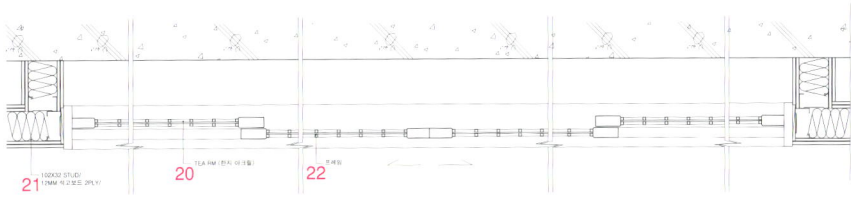

문 단면 상세 V / door section detail V

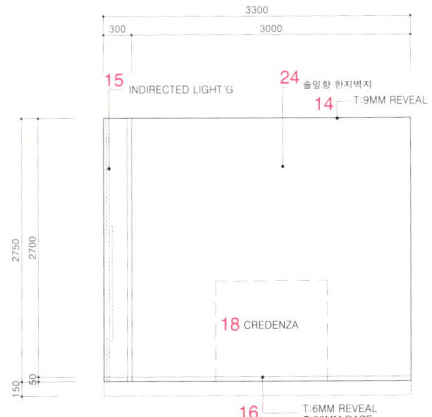

다실 입면 L / tea room elevation L

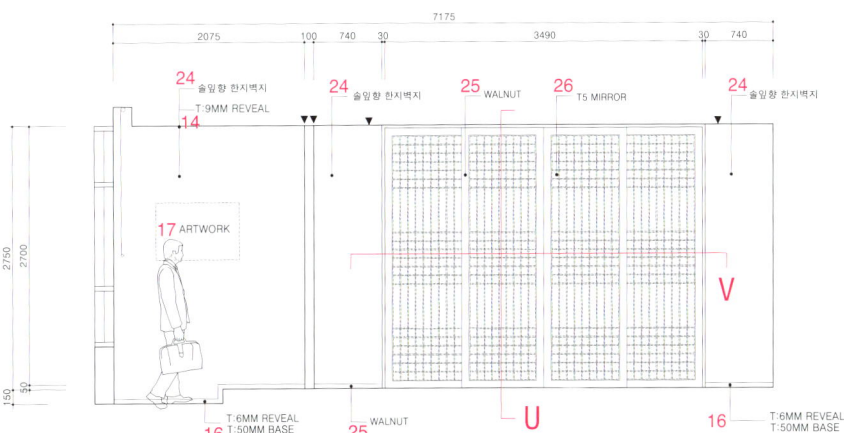

다실 입면 M / tea room elevation M

ING SECURITIES BROKERAGE CO., LTD.

(주)다원디자인 | 조서윤, 노경옥 Dawon Design Co., Ltd. | Fay Cho, Noh Kyung Ok

---

강남 파이낸스 센터 빌딩에 신규 설립하게 된 ING PB 증권 중개㈜는 글로벌 네트워크를 자랑하는 선진 금융업계의 선두주자이다. 이에 내부 인테리어는 직원들의 소속감 증대와 기업 이미지를 느낄 수 있도록 계획하였으며, 크게 3가지 공간(Client Area, Front Office, Back Office)으로 구성 되어 있어 유기적이면서 업무 능률향상을 이룰 수 있도록 중점을 두었다.

Client Area
Elevator Hall에서 Main 고객 출입구와의 동선이 긴 점을 감안하여 Clear Glass 자동문을 설치 하였으며 Sign Wall의 벽면을 Beige Tone Stucco Paint로 처리하여 Main Signage와 리셉션 데스크의 인지성을 높게 하였다. 특히 리셉션 공간은 중후하고 깔끔한 이미지를 고객에게 강조 할 수 있도록 한국적인 디자인 Motive(매,난,국,죽, 거친 스킨의 Wallnut Wood Flooring, 전통 한식 그릴의 Mordern화)를 사용하여 정적인 선들과 재질을 표한 할 수 있도록 하였다.

Front & Back Office
사무 공간의 벽면은 White, Maple Wood & Accent Orange Color로 처리하여 각 업무 공간의 독립성을 유지하면서 직원들을 위한 서비스 공간을 배치하여 편안함을 제공하는데 주력 하였다.

ING PB Stock Agency, Inc. established in Gangnam Finance Center building is the leader of the superior financial business, flaunting its global network. With this fact in mind, we planned the interior in which the company's image could be felt and employees could grow a sense of belonging. Divided into 3 spaces of Client Area, Front Office and Back Office, the place was designed centering on organized and efficient job performance.

Client Area
Taking into account the fact that a moving path from elevator hall to main entrance is long, an automatic door is installed, and we treated the surface of the sign wall with beige-tone stucco painting for the main signage and reception desk to get more recognition. The Reception Area is expressed in static lines and materials, as well as Korean traditional design motif (Plum Blossom, Orchid, Chrysanthemum, Bamboo, rough skin of Walnut wood flooring and modernized traditional grill) so that a noble and neat image would appeal to customers.

Front & Back Office
Using white, maple wood and accent orange color on the walls in the offices, the Service Area is arranged to provide a relaxing feeling to employees, while maintaining independence between each work space.

위치 서울시 강남구 역삼동 737 강남 파이낸스 센터 13층
용도 업무
면적 125평
설계기간 2008. 5 ~ 2008. 7
공사기간 2008. 6. 30 ~ 2008. 8. 10
공사담당 이석호
시공 다원디자인
마감 바닥 – 카펫, 우드 플로링 / 벽 – 안티코 스터코 페인트, 수퍼 그래픽, 천, 월 커버링 / 천정 – 페인트, 기성 천정 텍스타일
사진 이기환, 염승훈

Location 13th, GFC B/D, 737, Yeoksam-dong, Gangnam-gu, Seoul, Korea
Use Office
Area 125py
Design period 2008. 5 ~ 2008. 7
Construction period 2008. 6. 30 ~ 2008. 8. 10
Construction manager Lee Suk Ho
Construction Dawon Design Co., Ltd.
Finishing Floor - Carpet, Eood flooring / Wall - Anticco stuccoe paint, Super graphic, Fabric, Wall covering / Ceiling - Paint, Existing ceiling tex tile
Photographer Lee Ki Hwan, Yum Seung Hoon

천정도 / ceiling plan

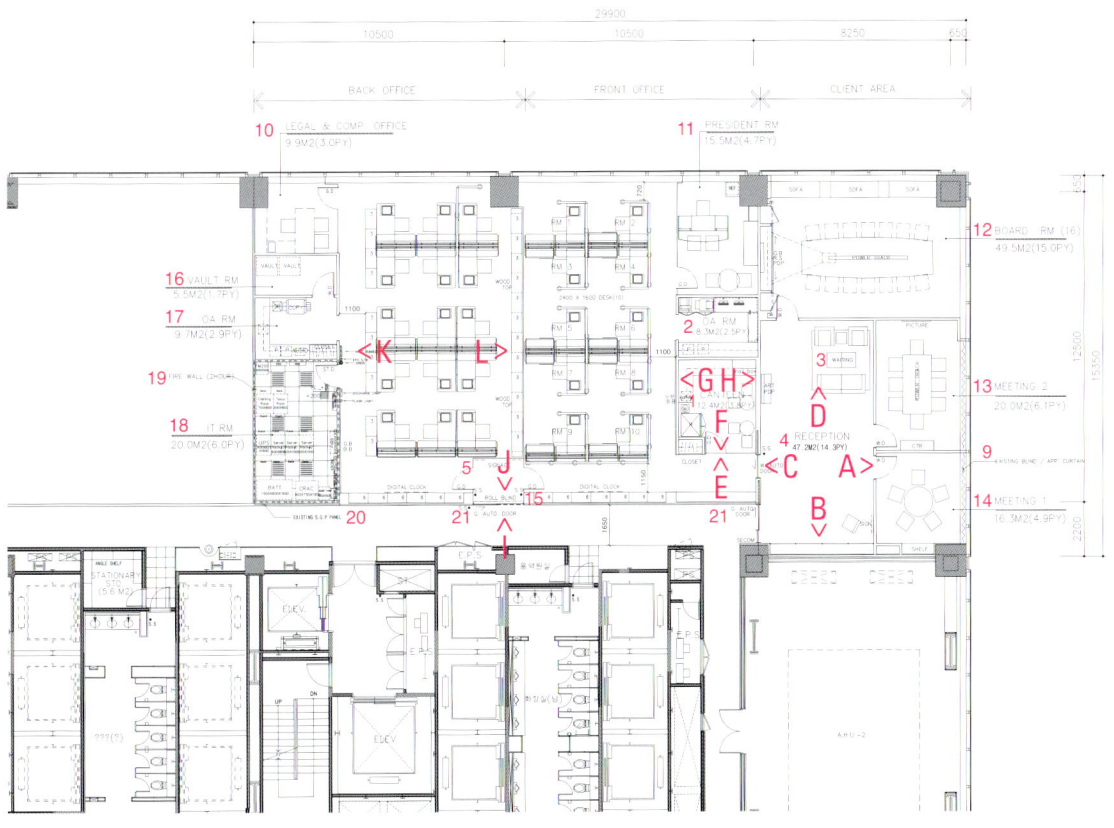

평면도 / floor plan

1 CANTEEN 12.4㎡(2.5PY) 2 OA RM. 8.3㎡(2.5PY) 3 WAITING 4 RECEPTION 47.2㎡(14.3PY) 5 SIGNAGE 6 APP. LED LIGHTING(ORANGE) 7 INDIRECT LIGHT 8 30X30 REVEAL 9 EXISTING BLIND, APP. CURTAIN 10 LEGAL & COMP. OFFICE 9.9㎡(3.0PY) 11 PRESIDENT RM. 15.5㎡(4.7PY) 12 BOARD RM.(16) 49.5㎡(15.0PY) 13 MEETING 2 20.2㎡(6.1PY) 14 MEETING 1 16.3㎡ (4.9PY) 15 ROLL BLIND 16 VAULT RM. 5.5㎡(1.7PY) 17 OA RM. 9.7㎡(2.9PY) 18 IT RM. 20.0㎡(6.0PY) 19 FIRE WALL(2HOUR) 20 EXISTING S.G.P PANEL 21 G. AUTO DOOR

1 탕비실 12.4㎡(2.5평) 2 OA실 8.3㎡(2.5평) 3 대기공간 4 리셉션 47.2㎡(14.3평) 5 사인물 6 지정 LED 조명(오렌지) 7 간접조명 8 30X30 창틀 9 기존 블라인드, 지정 커튼 10 법률 & COMP. 사무실 9.9㎡(3.0평) 11 대표자실 15.5㎡(4.7평) 12 보드실(16) 49.5㎡(15.0평) 13 회의실 2 20.2㎡(6.1평) 14 회의실 1 16.3㎡(4.9평) 15 롤 블라인드 16 금고실 5.5㎡(1.7평) 17 OA실 9.7㎡(2.9평) 18 IT실 20.0㎡(6.0평) 19 방화벽(2시간) 20 기존 S.G.P 패널 21 G. 자동문

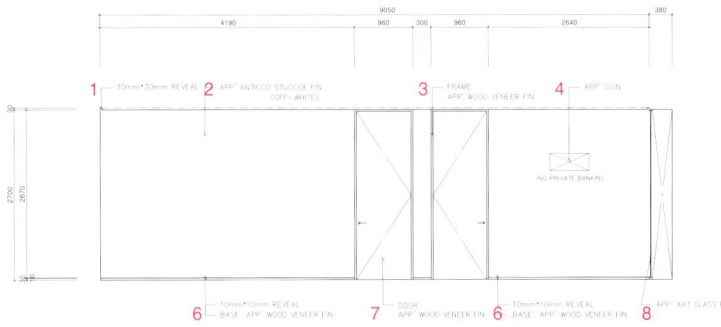

리셉션 입면 A / reception elevation A

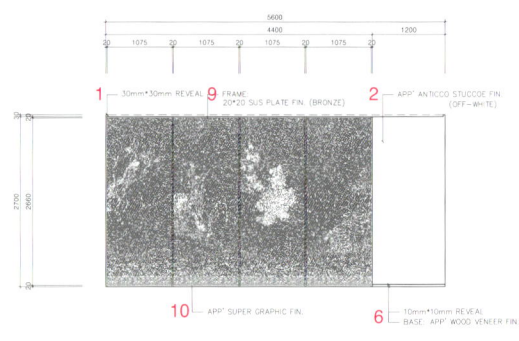

리셉션 입면 B / reception elevation B

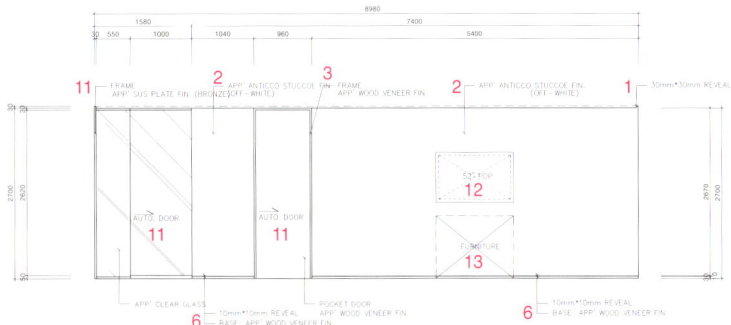

리셉션 입면 C / reception elevation C

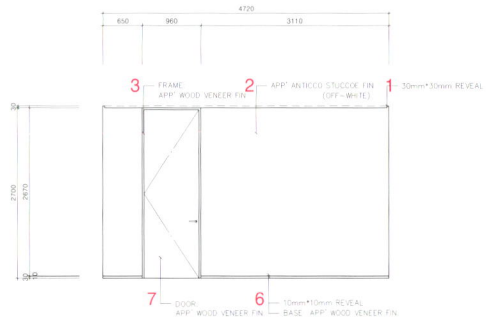

리셉션 입면 D / reception elevation D

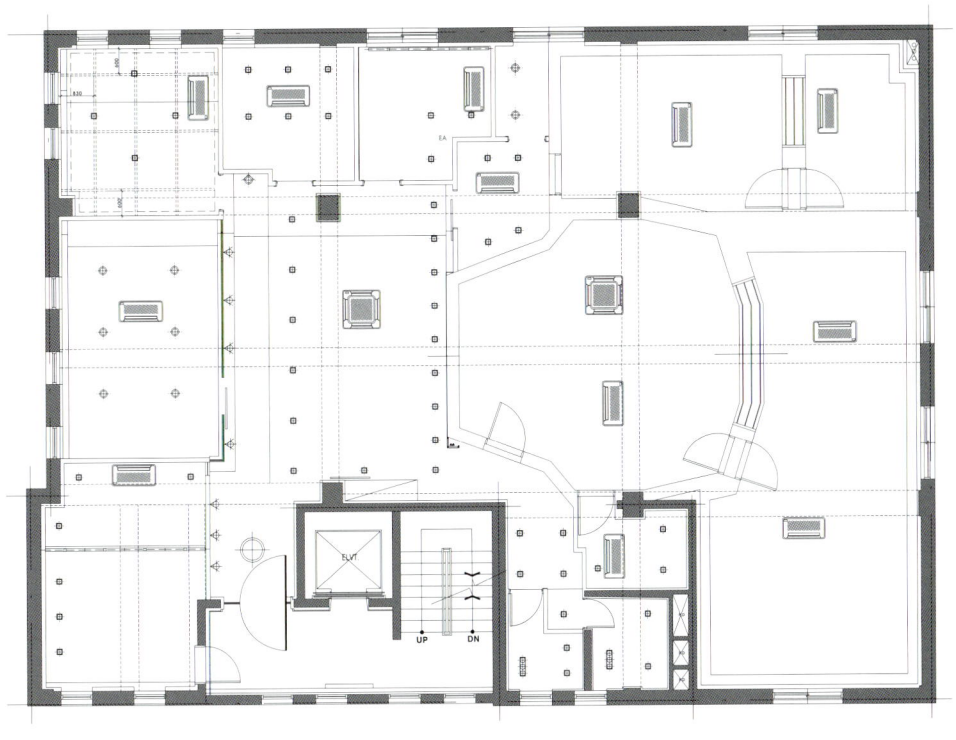

천정도 / ceiling plan

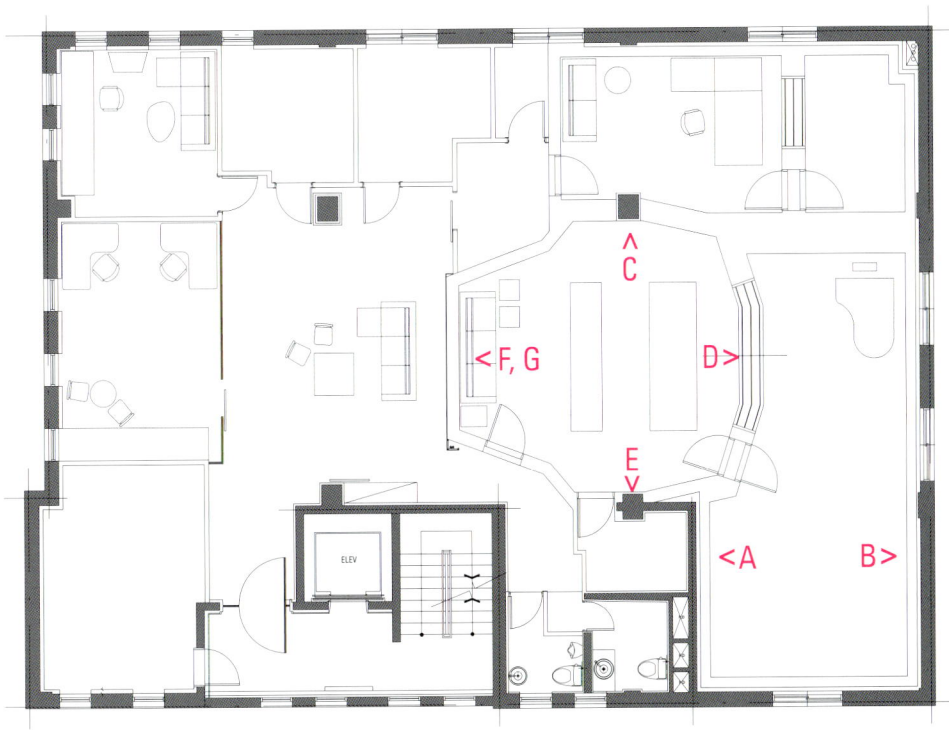

평면도 / floor plan

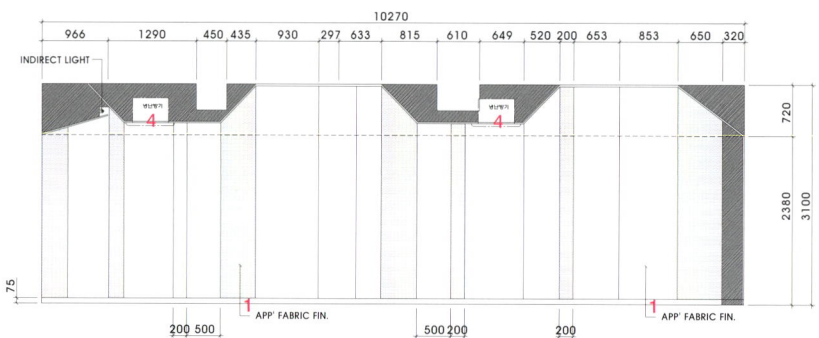

녹음 부스 입면 A / recording booth elevation A

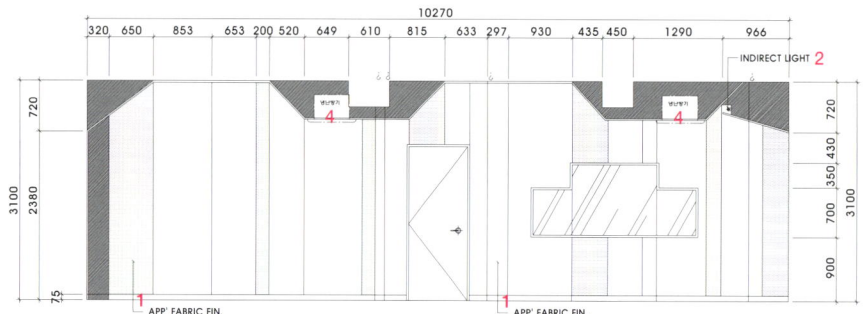

녹음 부스 입면 B / recording booth elevation B

1 APP. FABRIC FIN.　2 INDIRECT LIGHT　3 APP. WOOD FLOORING FIN.　4 COOLING AND HEATING MACHINE

1 지정 패브릭 마감　2 간접 조명　3 지정 우드 플로링 마감　4 냉난방기

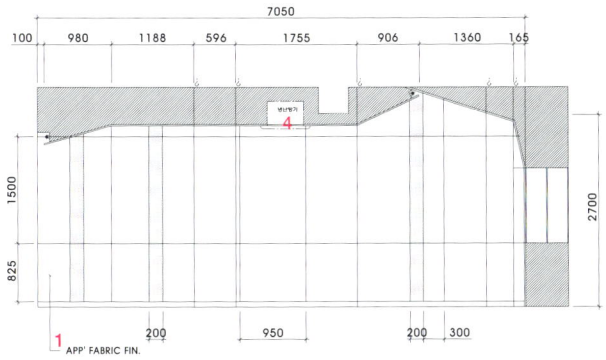

컨트롤실 입면 C / control room elevation C

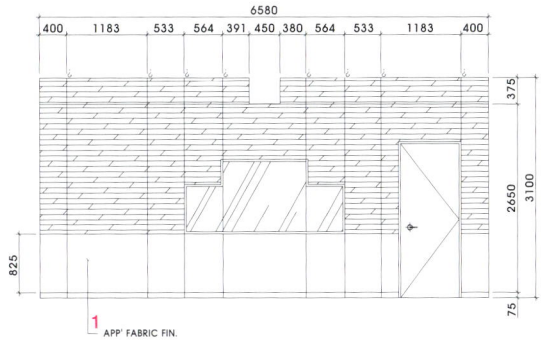

컨트롤실 입면 D / control room elevation D

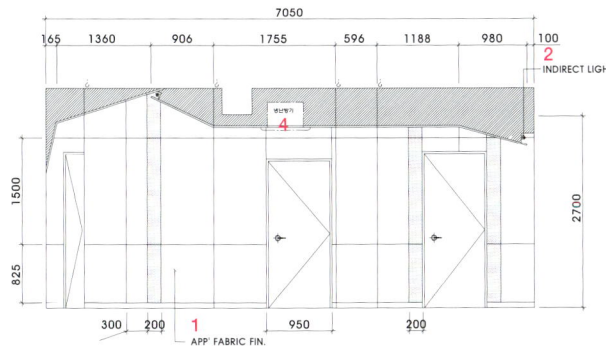

컨트롤실 입면 E / control room elevation E

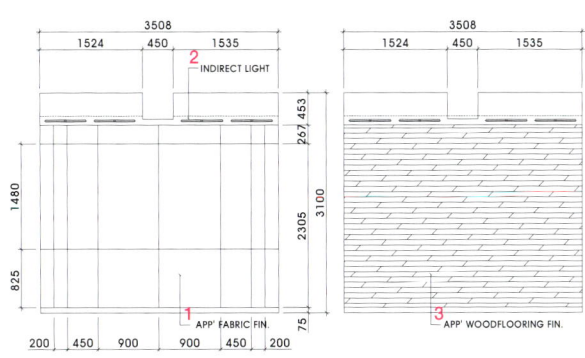

컨트롤실 입면 F, G / control room elevation F, G

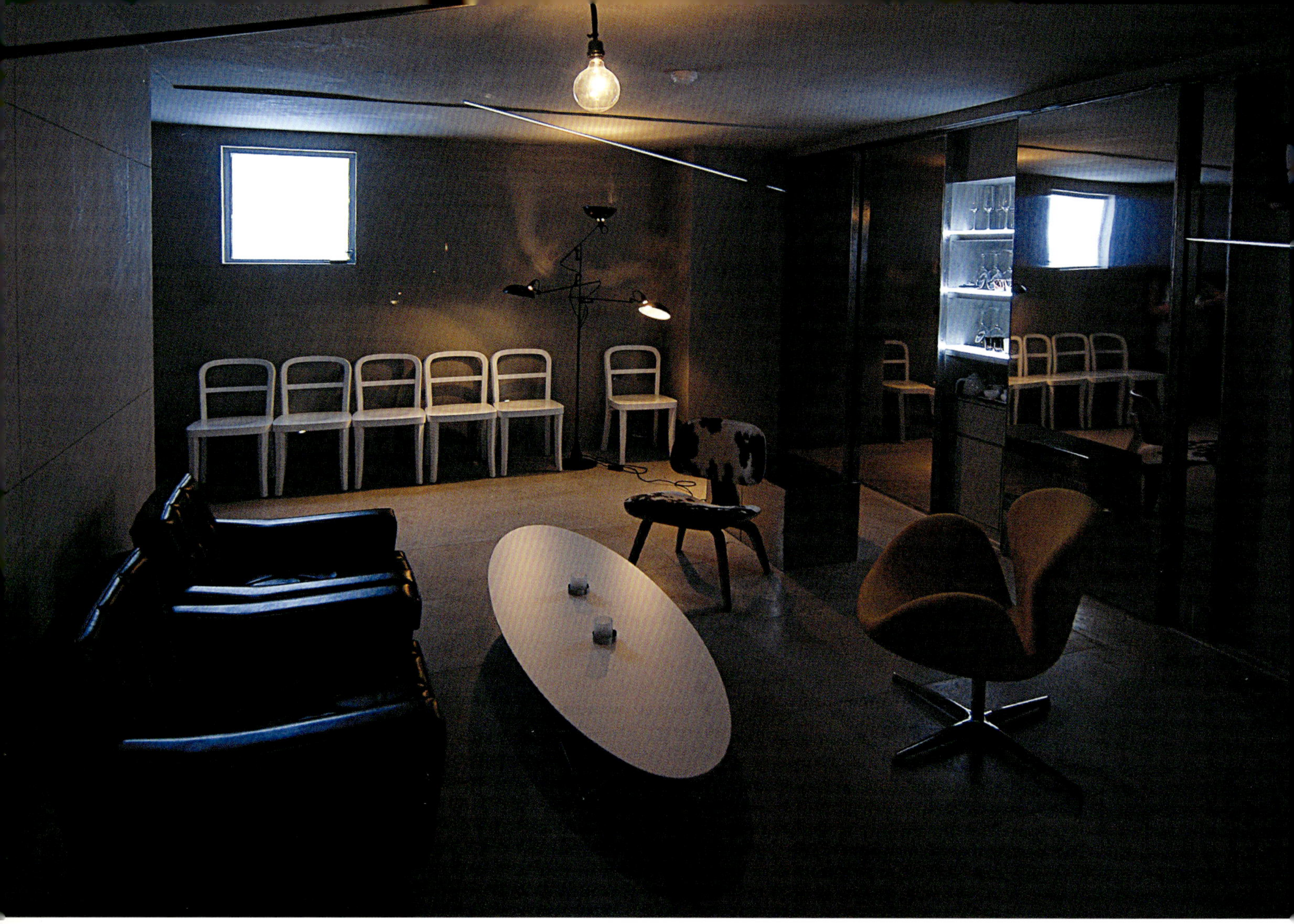

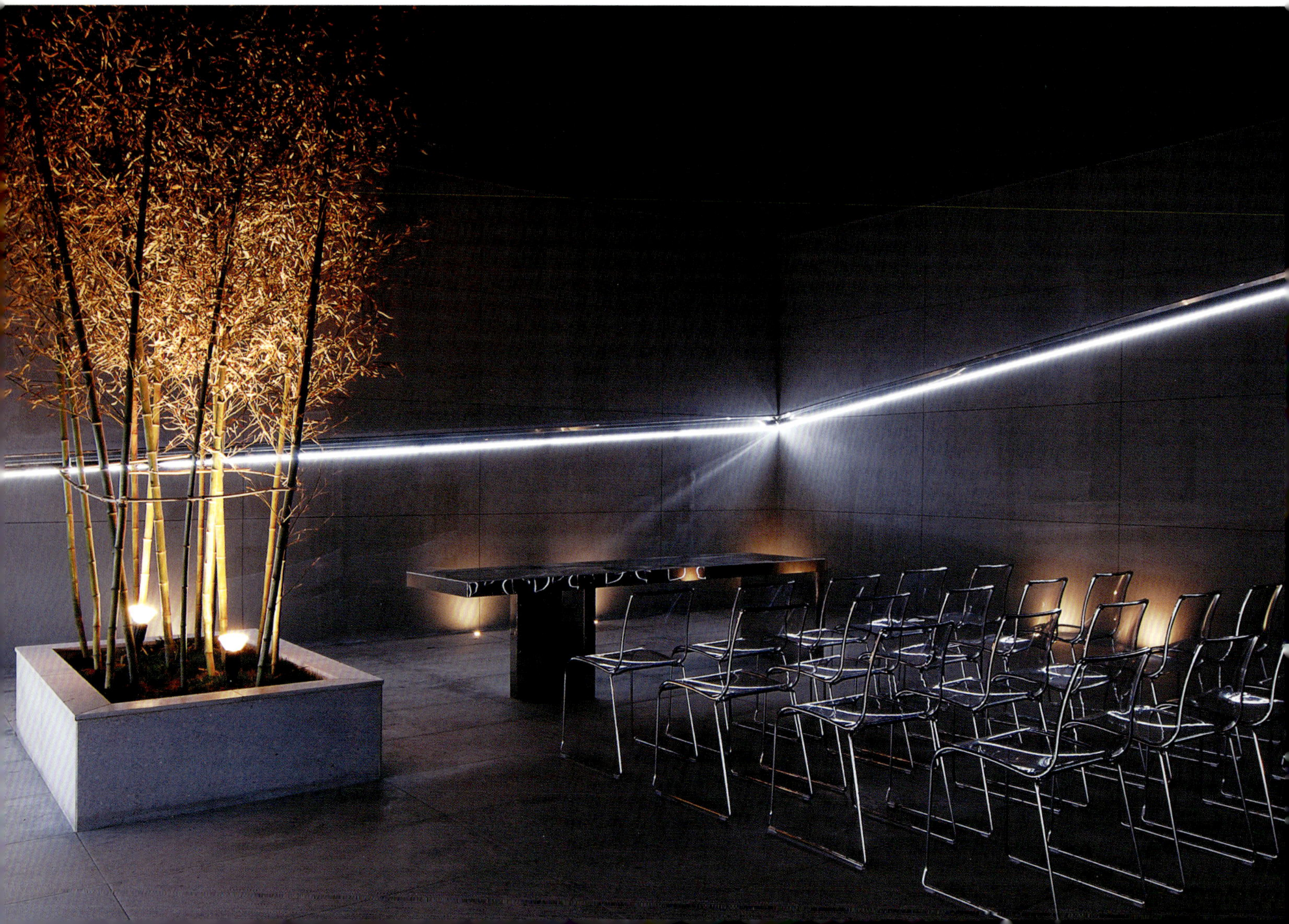

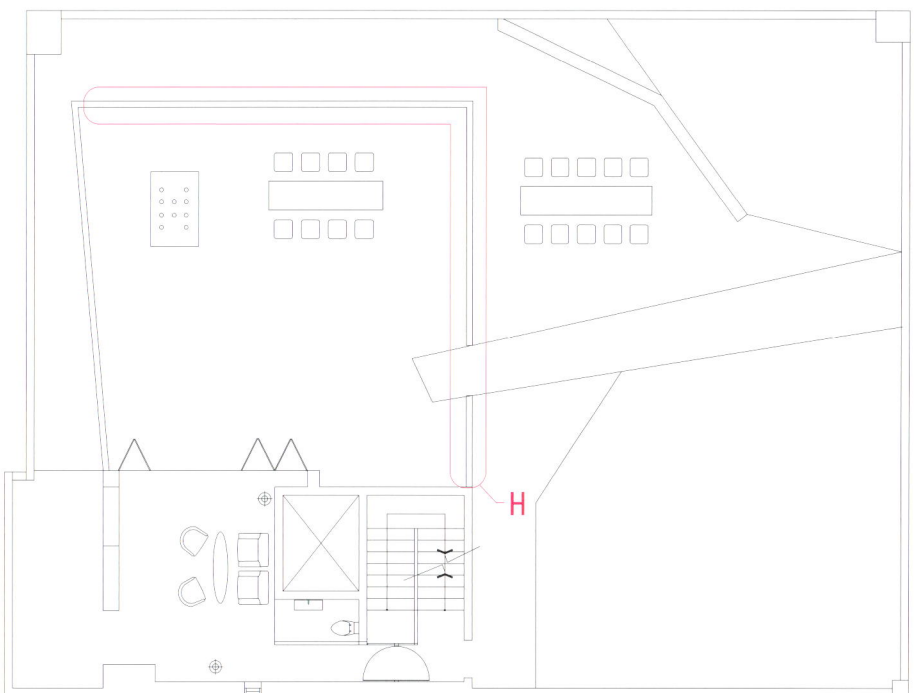

옥상 바 평면도 / roof floor bar plan

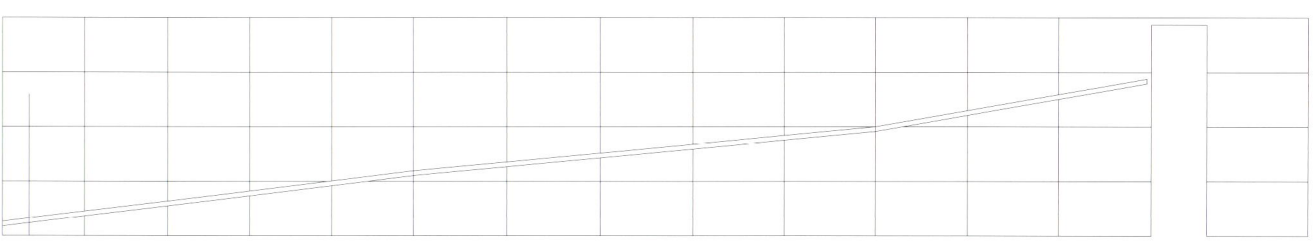

옥상 벽 입면 H / roof floor wall elevation H

DAIN LANDSCAPE OFFICE

(주) 이우진어소시에이트 | 이우진 LEEWOOJIN ASSOCIATE.CO.,LTD | Lee Woo Jin

사람과 나무, 休

프로젝트를 진행하면서 끊임없이 머릿속에 머물던 심상이었다. 자연을 언제나 가까이 곁에 두고 보고 싶어하는 소박한 욕심이 바로 조경 업(業)의 뜻이며, 동시에 이번 인테리어 방향이기도 했다. 자연을 통해 인간의 내부에서 절대적이고 순수한 본질을 이끌어내어 향유하고자 했던 뜻을 이 공간에서 재현해내고 하였다. 일하는 공간과 편안한 휴식의 공간이라는 공존할 수 없을 것 같은 두 개념이 조화된 Workplace인 동시에 Healing Place로서 존재하는 공간 말이다. 이는 자연주의-休로 이어졌다. 사람이 나무에 어깨를 기대 쉬고 있는 모습을 형상화한 休(휴)는 자연에서의 쉼인 동시에 인간과 나무, 곧 인간과 자연의 조화를 보여주고 있는 글자였다. 따라서 자연 및 조경이라는 심볼로서 나무를 인테리어의 테마로 자연을 간결한 형태로 형상화하였다. 마치 마을 초입의 큰 나무 그늘 아래 누운 것처럼 편안하고 포근한 느낌으로 누구나 꿈꾸는 편안한 자연의 느낌을 가질 수 있는 공간! 자연을 만나고 사람은 또 자연스러워지는 것처럼 이는 다인조경이 추구하는 뜻이자 공간을 통해 보여줄 것이다.

Human and Tree, 'Hyu(休)'

It was what I had been constantly thinking about while processing the project. The real meaning of landscape is a simple greed to put and look at nature next to me all the time. And it was also the interior direction this time. The intension to pull out a pure and absolute essence from inside the human and to enjoy it was represented in this space, in which two conflicted concepts of space were harmonized and co-existed as Workplace and Healing place. This continued with Naturalism, 'Hyu(休)'. As a Chinese character symbolizing the look of a person leaning his shoulder against a tree, 'Hyu(休)' shows the concept of relaxing in nature and the harmony between human and tree or nature at the same time. Therefore, as the symbol of nature and landscape, a tree was used for the interior theme to embody nature in a simple shape. As a space where comfortable and snug feelings can be shared as if you are lying down under the shadow of a big tree standing in the entrance of village, "Meeting nature and Being natural" is what Dain Landscape pursues and shows through the space.

위치 경기도 수원시 영통 신동 486번지 102-615
용도 업무
면적 205.29㎡
설계기간 2008. 6. 1 ~ 2008. 6. 20
시공기간 2008. 7. 1 ~ 2008. 7. 31
실시설계 (주)이우진어소시에이트 / 이진희, 전민관
시공 (주)이우진어소시에이트 / 전민관, 전재경
건축주 이승용
마감 바닥 - P-타일, 카펫 / 벽 - 칼라 락카 도장, 투명 락카, 자작 합판 / 천정 - 비닐 페인트
사진 (주)이우진어소시에이트 제공

Location 486 Sin-dong, Yeongtong-gu, Suwon-si, Gyeonggi-do, Korea
Use Office
Area 205.29㎡
Design period 2008. 6. 1 ~ 2008. 6. 20
Construction period 2008. 7. 1~ 2008. 7. 31
Project Design LEEWOOJIN ASSOCIATE.CO.,LTD / Lee Jin Hee, Jun Min Kwan
Constrution LEEWOOJIN ASSOCIATE.CO.,LTD / Jun Min Kwan, Jun Jae Kyung
Client Lee Seung Yong
Finishing Floor - P-Tile, Carpet / Wall - Color lacquer painting, Clear lacquer, Birch plywood / Ceiling - V.P
Photos offer LEEWOOJIN ASSOCIATE.CO.,LTD

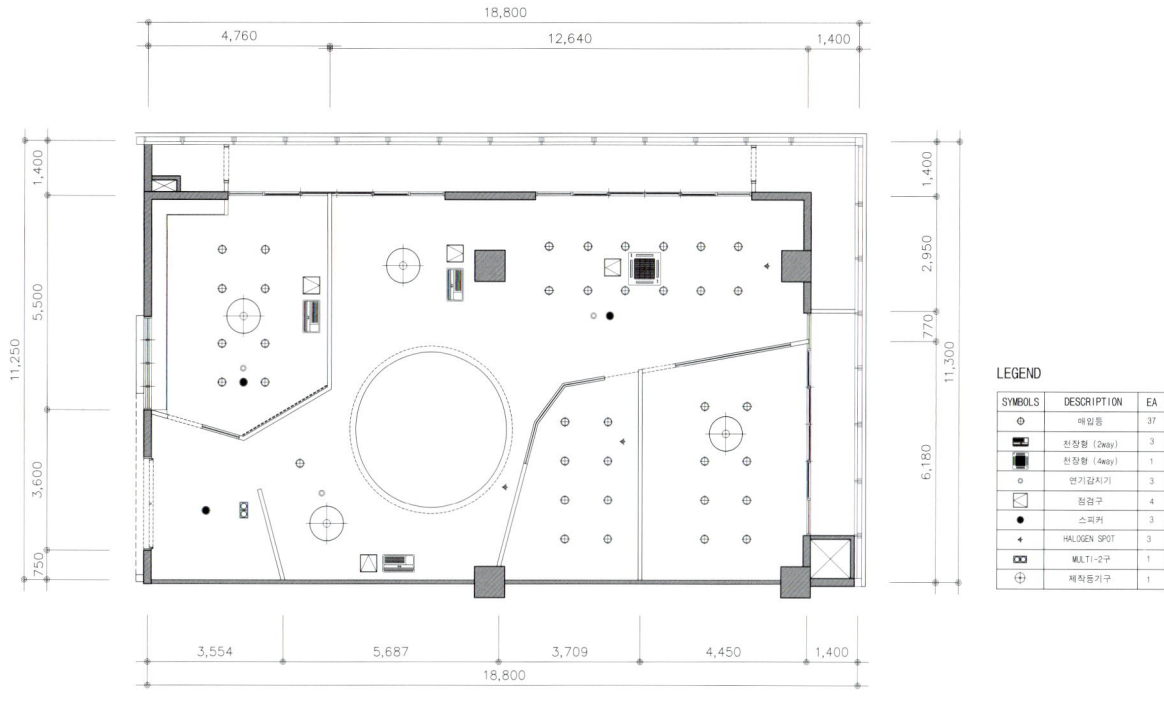

천정도 / ceiling plan

평면도 / floor plan

1 ENTRANCE 2 CONFERENCE ROOM 3 IMAGE WALL 4 A TEAM 5 B TEAM 6 ADMINISTRATION TEAM 7 DIRECTOR ROOM 8 PRESIDENT ROOM 9 BAR 10 STORAGE

1 입구 2 회의실 3 이미지 월 4 A팀 5 B팀 6 관리팀 7 디렉터 실 8 대표실 9 바 10 창고

1 INDIRECT LIGHTING DECO 28W **2** T18MM MDF ON WHITE COLOR LACQ FIN. **3** APP. T8MM TEMPERED GLASS FIN. / APP. SUS FIN. **4** T18MM BIRCH PLYWOOD ON CLEAR LACQ FIN. **5** APP. T8MM TEMPERED GLASS FIN. / APP. LIGHT GREEN COLOR LACQ FIN. **6** APP. MDF LIGHT GREEN LACQ FIN. **7** APP. T18 BIRCH PLYWOOD ON CLEAR LACQ FIN. **8** APP. T1.6MM GALVA ON WHITE POWDER PAINTING FIN. **9** □38×38 SQUARED LUMBER **10** T18MM MDF **11** INDIRECT LIGHTING DECO 28W(T5) **12** APP. T8 TEMPERED GLASS ON SHEET FIN./ APP. SUS FIN.

DAIN LANDSCAPE OFFICE | 다인 조경 오피스

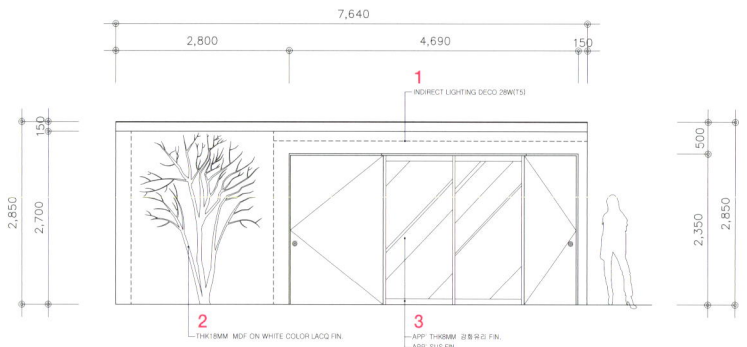

입구 입면 A / entrance elevation A

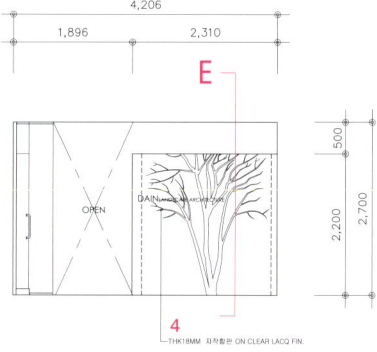

이미지 월 입면 B / image wall elevation B

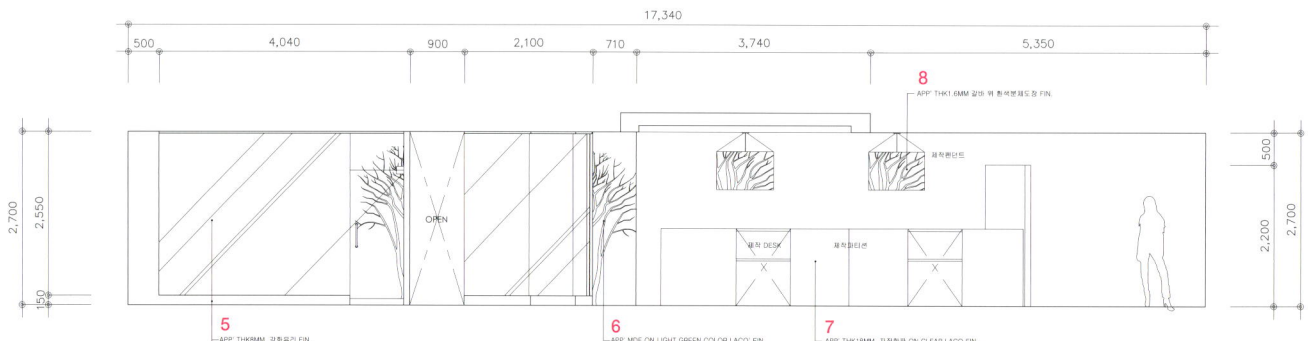

입면 C / elevation C

1 간접 조명 데코 28w 2 T18mm MDF 위 흰색 락카 마감 3 지정 T8mm 강화유리 마감/ 지정 서스 마감 4 T18mm 자작합판 위 투명 락카 마감 5 지정 T8mm 강화유리 마감 / 지정 연두색 락카 마감
6 지정 MDF 연두색 락카 마감 7 지정 T18 자작합판 위 투명 락카 마감 8 지정 T1.6mm 갈바 위 흰색 분체도장 마감 9 ㅁ-38×38 각재 10 T18mm MDF 11 간접 조명 데코 28W(T5) 12 지정 T8 강화유리 위 시트 마감/ 지정 서스 마감

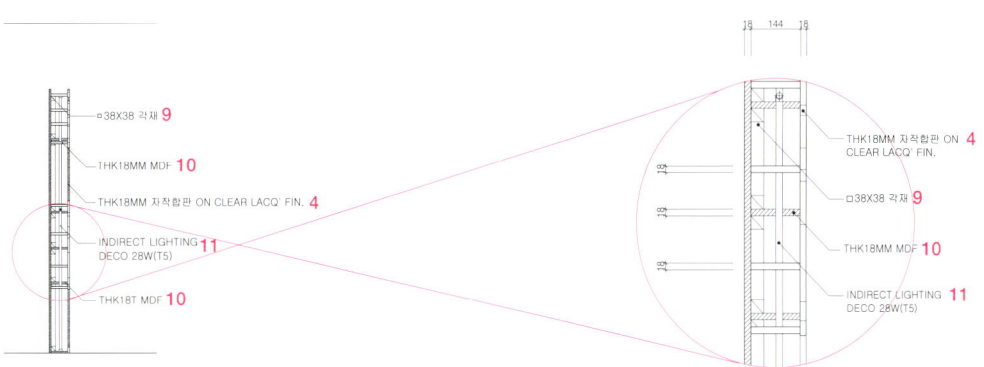

이미지 월 단면 상세 E / image wall section E

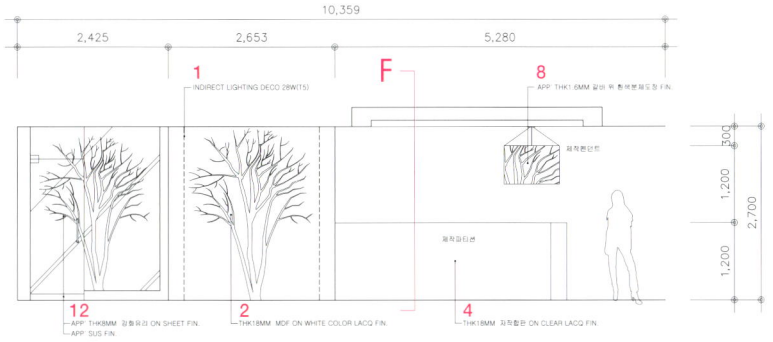

입면 D / elevation D

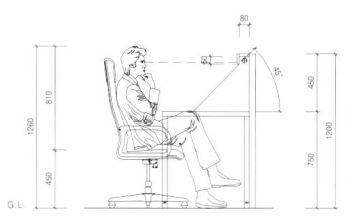

제작 파티션&책상 상세 F / producted partition&desk detail F

ALPHA VILLE 44 PRODUCTION

스튜디오 베이스 | 전범진　Studio VASE l Jun Bum Jin

장 릭 고다르는 영화사 최고 거장으로 추앙받고 있다. 그가 남긴 영화 알파빌(Alphaville, 1965)의 제목을 따 붙인 곳으로 세간에 회자되는 cf를 만드는 회사이다. 시대 흐름의 최전선에서, 극도로 감각적이어야 할 사람들이 머무르는 광고 회사인 것이다. 친분에 의해 자연스럽게 시작된 작업은 그간 알지 못했던 내재된 나의 다른 면모를 발견할 수 있는 기회였다. 알파빌44프로덕션은 자연스러운 미니멀스타일이다. 내부의 낡은 벽돌로 이뤄진 벽은 뉴욕에서 공부해 빈티지 스타일의 첼시 지구를 좋아했던 클라이언트의 취향을 반영했다. 1층 입구의 벽면을 안티코 스터코로 마감한 뒤, 이를 스크린으로 삼아 프로젝트 영상을 쐈다. 이 영상은 그냥 미키마우스가 아닌 1928년 제작된 "vintage mickey"란 영화이다. 지금 우리가 보아도 정말 신나게 웃으며 볼 수 있는 만화로 아주 오래된 역사를 지닌 것이지만 현대에 와서도 이 폼(form)을 답습하고 있을 정도 이니 이런 것이 결국 본질이 아닌가 생각한다. 1, 2층과 달리 높은 (4000m) 천장고의 3층은 두개의 회의실로 두었고, 작은 파티도 겸할 수 있는 공간이다. 공간디자인에도 내러티브가 있어야 한다는 생각에 동의하듯 3층까지 올라오며 외부 방문객이 회사를 둘러보는 동안의 시나리오와 3층 입구에 있는 터널을 지날때 유리벽 너머로 회의실을 바라보는 시퀀스를 예상하며, 이곳에서 사람들이 느낄 '심리적 공간'을 염두해 두었다.

Jean Luc Godar has been in high esteem as the greatest film maker in the history. The company building was named after one of his movies 'Alphaville(1965)'. It is a space for people who make commercials, thus the space had to be decorated as hip as their job. The project provided me a worthwhile opportunity to find myself in me. Alphaville 44 production is in natural minimal style.
The worn-out bricks of the inner walls reflect unique taste of the client who likes vintage since he studied in New York. After finishing the wall at the entrance with antico stucco, I projected images on the wall as screen. The film is a movie titled "vintage mickey" which was made in 1928. Even nowadays, we can truly enjoy the classical mickey mouse movie. In the end, we have followed the old form and I think that is the essence. Unlikely the first floor nor the second floor, the 4000m high third floor contains two meeting rooms where can also be used for a small party. As I believe that a space should be designed based on its own narrative, in designing the spaces, I drew a scenario that a visitor looks into the meeting room through the glass wall while ascending to the third floor.

위치 서울시 강남구 청담동 112-17번지
건축주 Alpha Ville 44 박성민
용도 업무
면적 330m²
설계기간 25일간
시공기간 45일간
마감 바닥-에폭시코팅 벽-V.P 천장- V.P
사진 박우진

Location 1112-17 Cheongdam-dong Gangnam-Gu Seoul. korea
Client Alpha Ville 44 Park Sung Min
Use Office
Finishing Floor-epoxy coating Wall-v.p Ceiling-v.p
Photographer Park Woo Jin

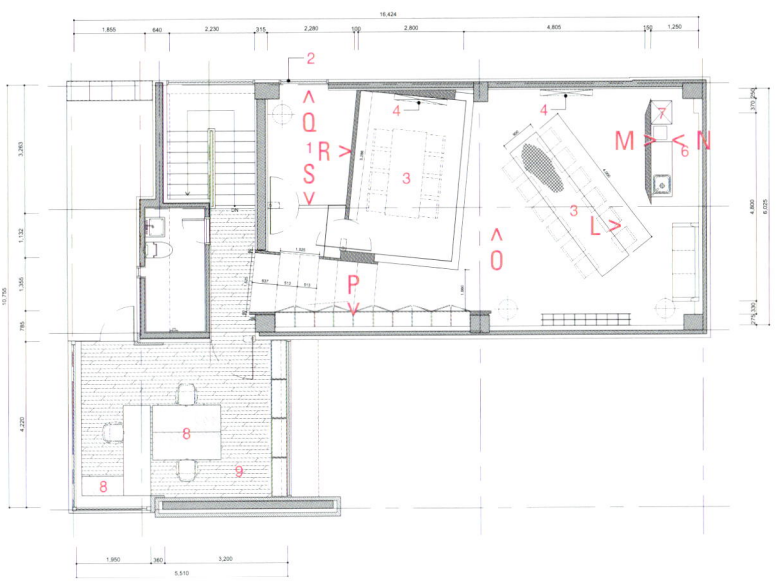

3층 평면도 / 3rd floor plan

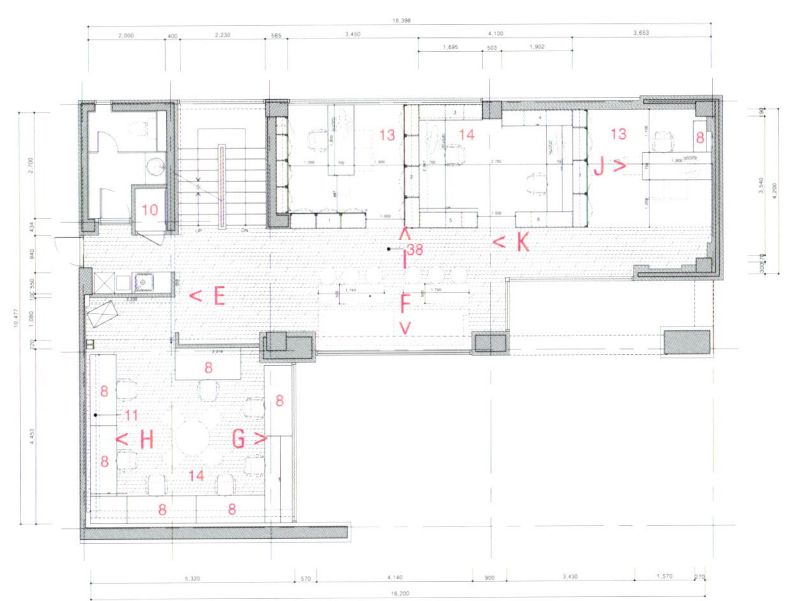

2층 평면도 / 2nd floor plan

1 president room 2 existing window opening and shutting type replacement 3 meeting room 4 PDP 5 conference room 6 canteen 7 refrigerator 8 existing furniture 9 planning room 10 electric room 11 upper video collection 12 meeting room 13 director room 14 assistant director room 15 toilet : reusable 16 edit machine 17 Floor : 500X500 white bianco tile fin. 18 electric water heater 19 removal sheet 20 Floor : 500X500 white glossy tile fin. 21 Fabric 22 Floor : existing floor on enamel paint fin. 23 movie team 24 A/C 25 Floor : urethane coating fin. 26 Lobby 27 deer's foot without-institution(metal check) 28 Mirror 29 Info Desk 30 Pendant 31 desk 32 management team 33 document room 34 door stopper check 35 mailbox paintwork 36 strongbox 37 removal painting 38 Floor : existing floor cut off / Mokyeom fin.

1 사장실 2 기존창호 개폐형으로 교체 확인 3 소회의실 4 PDP 5 대회의실 6 탕비실 7 냉장고 8 기존 가구 9 기획실 10 전기실 11 상부 비디오 수납 12 회의실 13 감독실 14 조감독실 15 화장실 : 재사용 16 편집기 17 바닥 : 500X500 화이트 비안코 타일 마감 18 전기온수기 19 쉬트 철거 20 바닥 : 500X500 화이트 유광 타일 마감 21 페브릭 22 바닥 : 기존바닥 위 애나멜 도장 마감 23 영화팀 24 A/C 25 바닥 : 우레탄 코팅 마감 26 로비 27 노루발 미설치(금속 확인) 28 거울 29 인포메이션 데스크 30 펜던트 31 책상 32 관리팀 33 문서실 34 도어스토퍼 확인 35 우체통 도색 36 금고 37 도장철거 38 바닥 : 기존 바닥 깍기 / 목염 마감

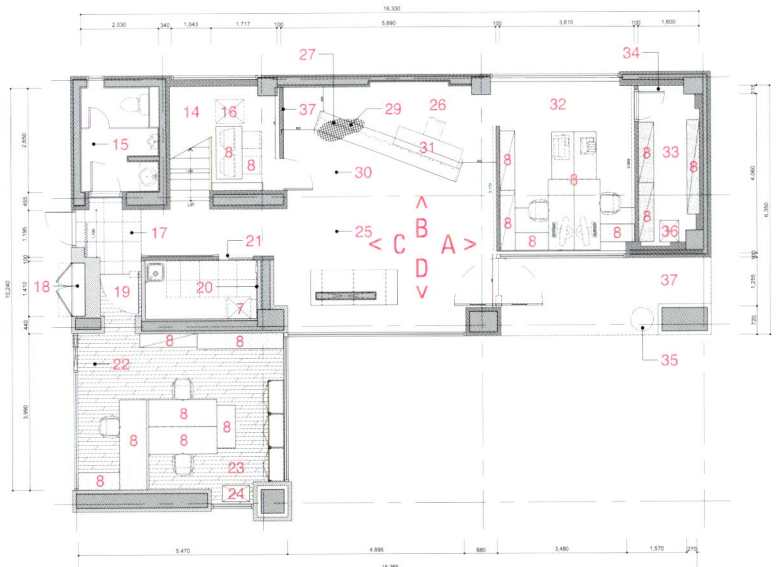

1층 평면도 / 1st floor plan

1 existing masonry wall 2 exposed ceiling on water paint fin. 3 App.lighting installing 4 Black SUS panel fin. 5 Lobby 6 Antico Stucco(White) fin. 7 toilet 8 staircase 9 Door : lacquering painting fin.(handle check) 10 editing room 11 gypsum board on V.P fin. 12 Door : Mirror SUS fin.(Design, finishing, handle check) 13 ENT 14 App. curtain installing 15 THK 5mm Mirror installing 16 Open 17 canteen 18 movie team 19 Door : THK 5mm Mirror installing(handle, hinge check) 20 outdoor

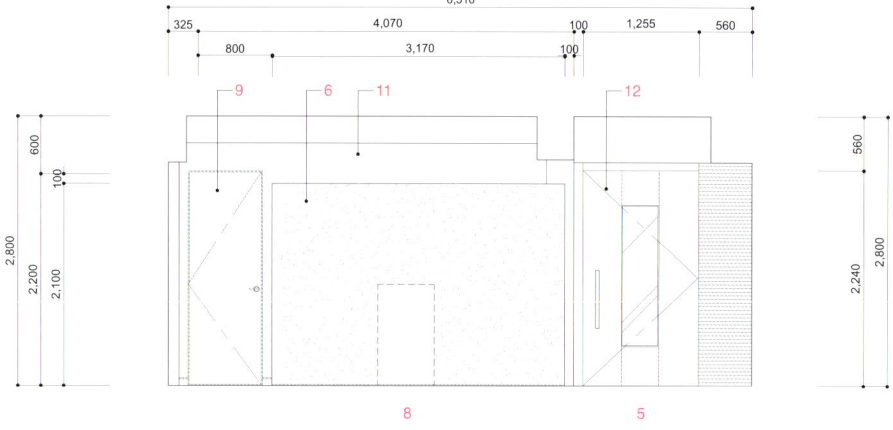

1층 인포메이션 로비 입면 A / 1st floor information lobby elevation A

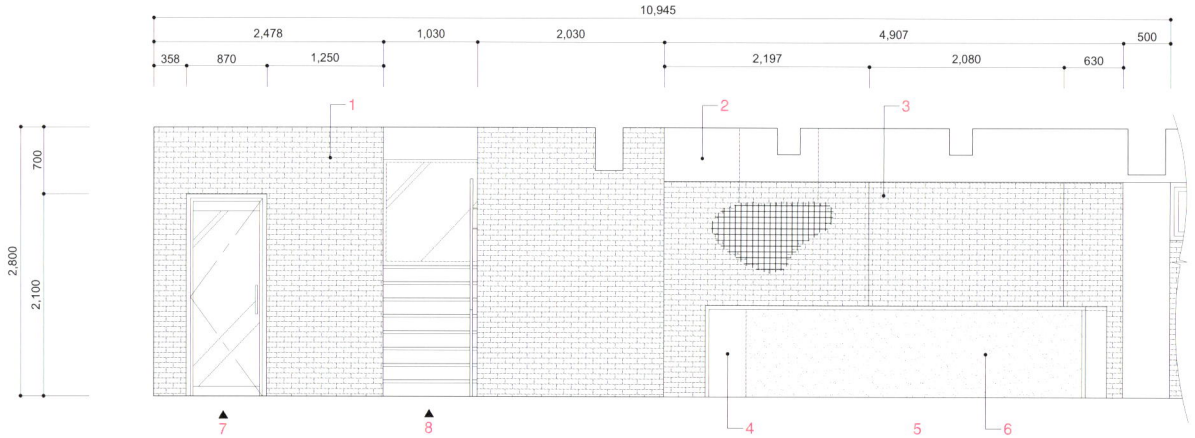

1층 인포메이션 로비 입면 B / 1st floor information lobby elevation B

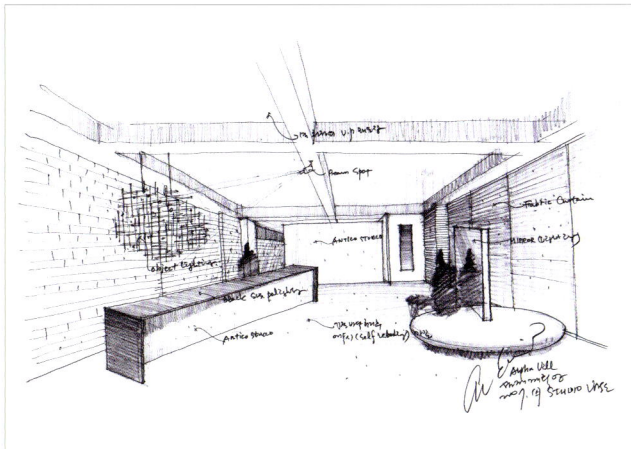

1 기존 조적 벽체 2 노출 천정 위 수성도장 마감 3 지정 조명 취부 4 Black SUS 패널 마감 5 로비 6 안티코 스터코(흰색) 마감 7 화장실 8 계단실 9 문 : 락카 도장 마감(손잡이 확인) 10 편집실 11 석고보드 위 V.P도장 마감 12 문 : 거울 SUS 마감(디자인, 마감재, 손잡이 확인) 13 ENT 14 지정 커튼 취부 15 THK 5mm 거울취부 16 오픈 17 탕비실 18 영화팀 19 문 : THK 5mm 거울 취부(손잡이, 힌지 확인) 20 외부

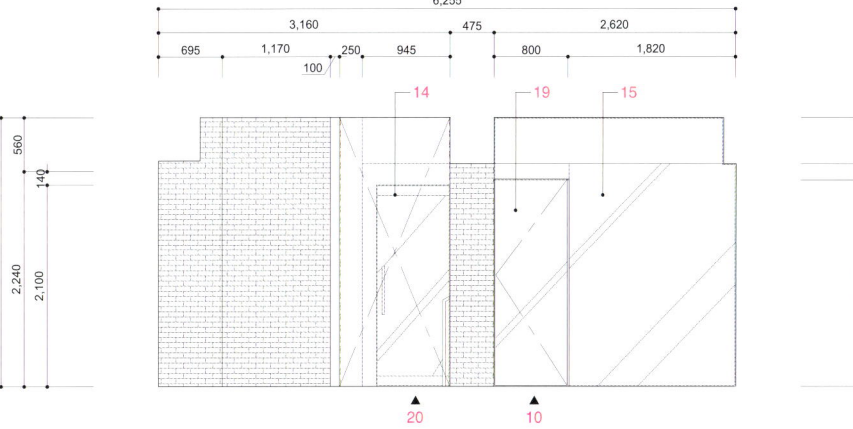

1층 인포메이션 로비 입면 C / 1st floor information lobby elevation C

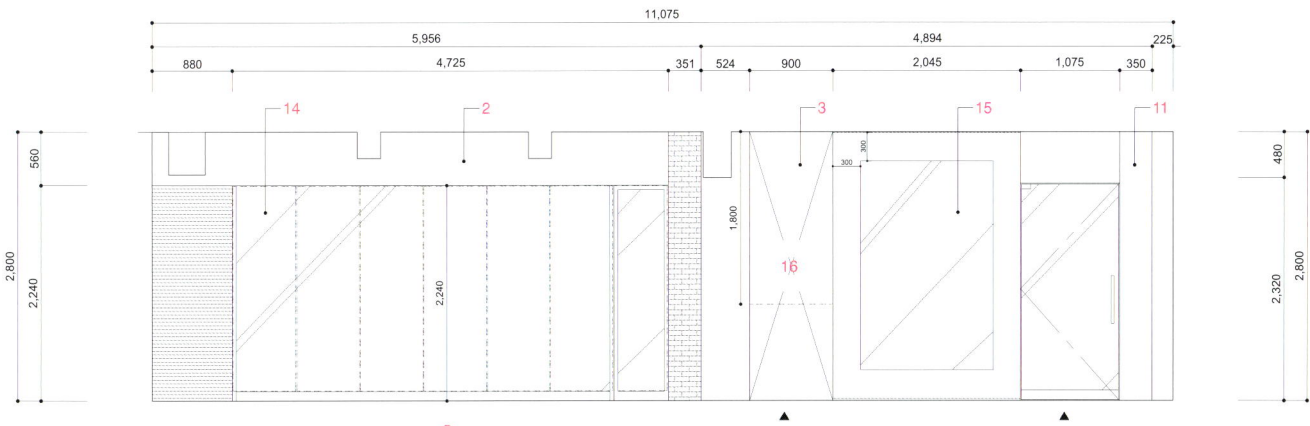

1층 인포메이션 로비 입면 D / 1st floor information lobby elevation D

1 V.P fin. 2 reflection glass installing 3 corridor 4 meeting room 5 App.veneer wood fin. 6 canteen 7 App. curtain installing 8 assistant director room 9 canteen, toilet 10 T5mm Mirror installing 11 T18mm MDF on inky water dyeing painting fin. / Door : lacquer painting fin. 12 gypsum board on V.P fin. 13 director's room 14 Sofa 15 open 16 corridol(meeting room) 17 T18mm MDF

1 V.P 마감 2 반사유리 취부 3 복도 4 회의실 5 지정 무늬목 마감 6 탕비실 7 지정 커튼 취부 8 조감독실 9 탕비실, 화장실 10 T5mm 거울 취부 11 T18mm MDF 위 먹물염색 도장 마감 / 문 : 락카 도장 마감 12 석고보드 위 V.P 도장 마감 13 감독실 14 소파 15 열린 공간 16 복도(회의실) 17 T18mm MDF

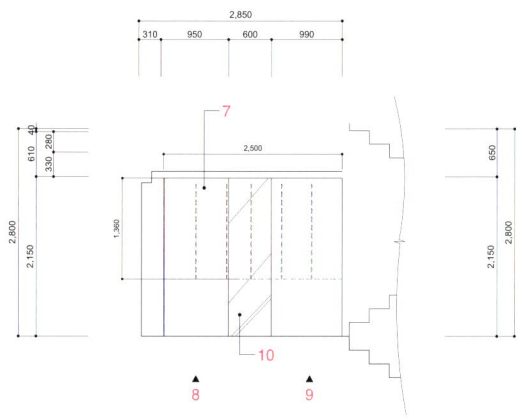

2층 사무공간 입면 E / 2nd floor office elevation E

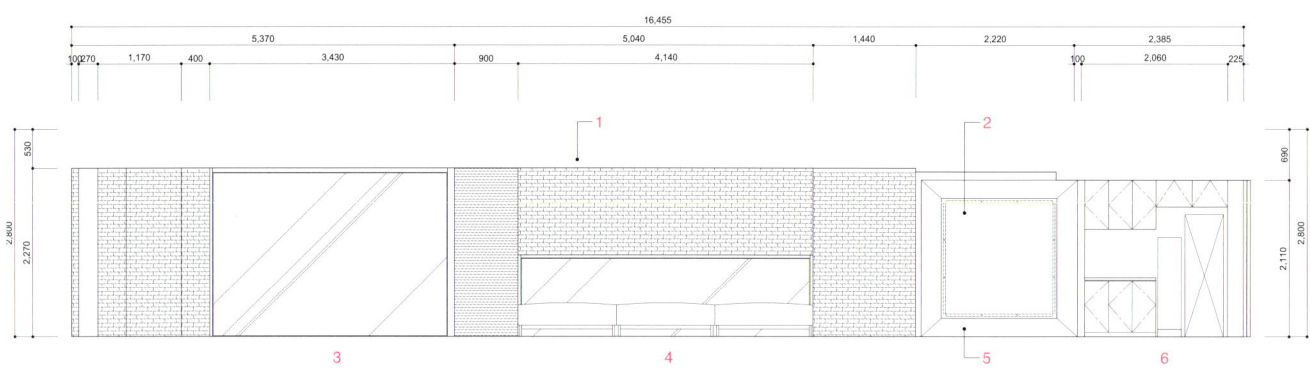

2층 사무공간 입면 F / 2nd floor office elevation F

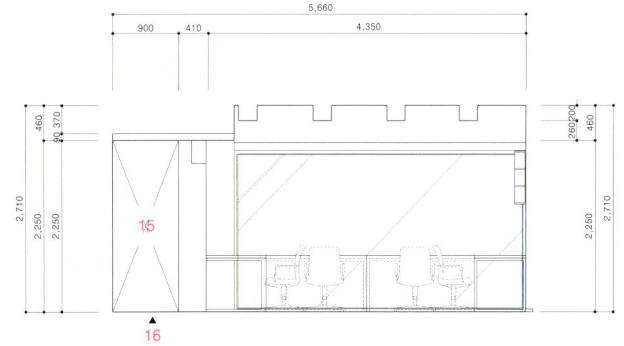

2층 조감독실 입면 G / 2nd floor assistant director's room elevation G

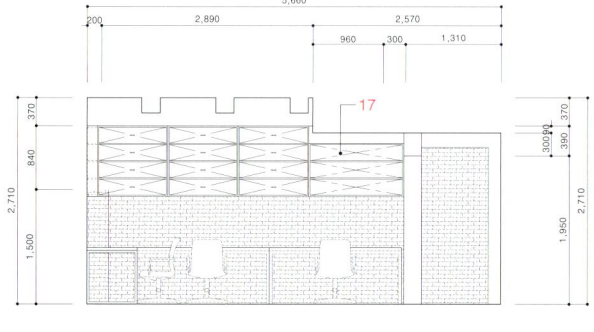

2층 조감독실 입면 H / 2nd floor assistant director's room elevation H

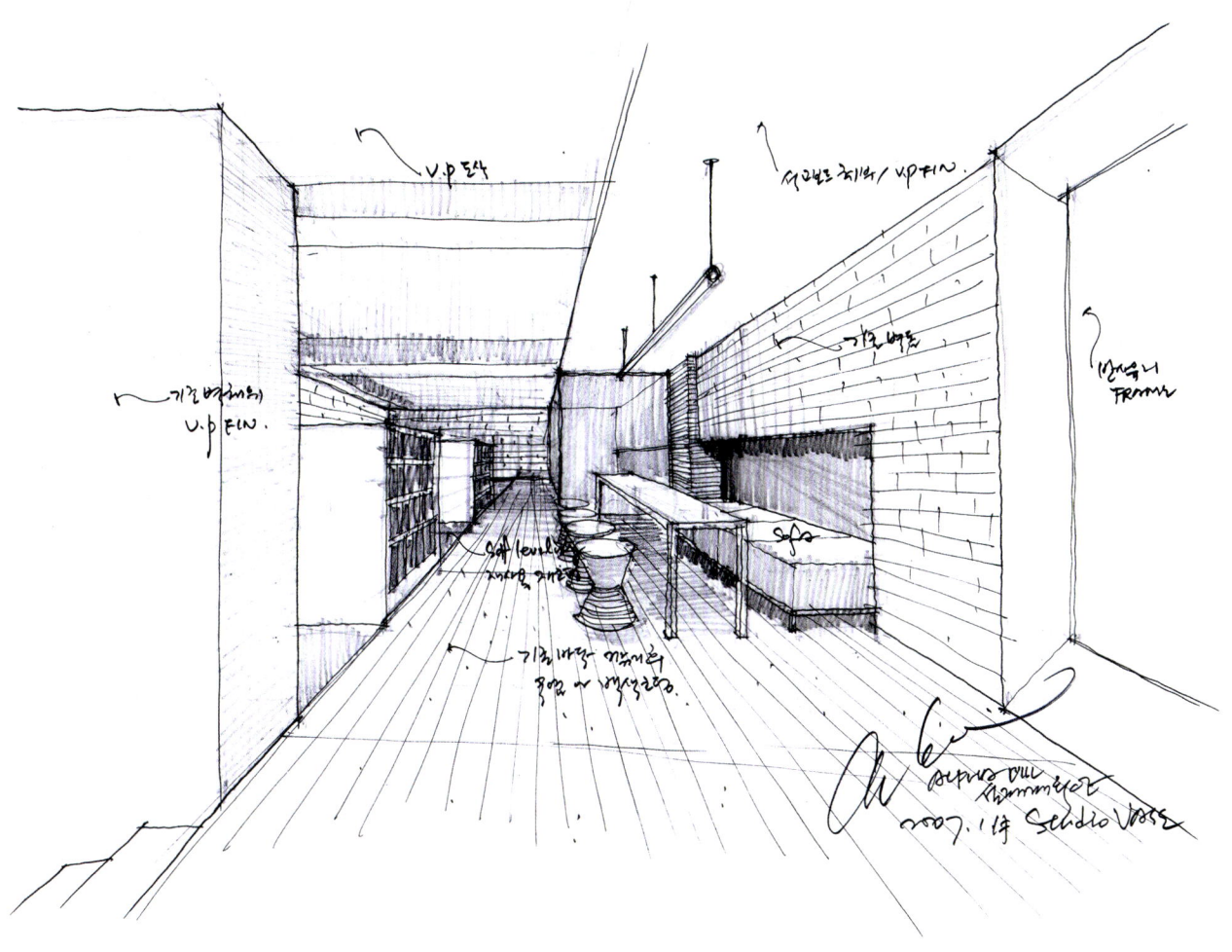

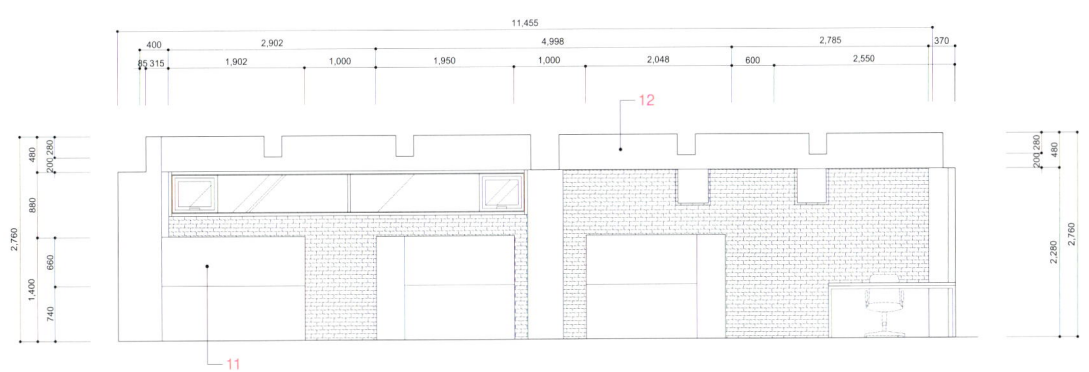

2층 사무공간 입면 I / 2nd floor office elevation I

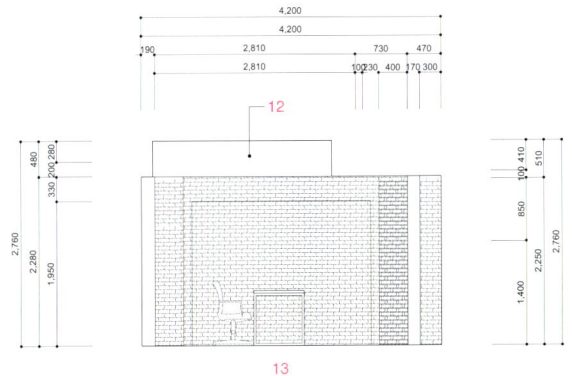

2층 사무공간 입면 J / 2nd floor office elevation J

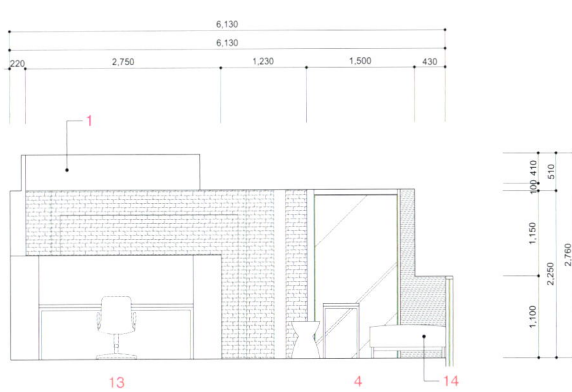

2층 사무공간 입면 K / 2nd floor office elevation K

1 V.P fin. 2 conference room 3 Open 4 Antico Stucco(White) fin. 5 canteen 6 shelf : pipe reinforcing , painting fin. 7 galva on antico stucco fin. 8 galva on antico stucco fin. / 10mm joint 9 galva on White Sheet fin. / shelf : pipe reinforcing , painting fin. 10 refrigerator 11 small cymbals 12 ready-made furniture(White) 13 Hot rolling steel plate installing 14 Antico Stucco(Black) fin. 15 corridol 16 Hot rolling steel plate 17 mosquito net installing/window built type insulation 18 frame : galva on black painting fin. /smallest thickness. 19 clear glass on color sheet + frost sheet installing/project window(odosi)/mosquito net installing/window built type insulation 20 gypsum board on V.P fin./inner sound isolation(Styrofoam) 21 ceiling joint 10X10mm 22 tempered door : handle 23 base 10X10mm : oil paint fin. 24 stair front : gypsum board on V.P fin. 25 glass groove digging /clear glass on frost sheet installing 26 tempered door : handle, lower part device for locking 27 10X10mm sus(mirror)glass groove

1 V.P 마감 2 대회의실 3 열린 공간 4 안티 스터코(흰색)마감 5 탕비실 6 선반 : 파이프 보강 , 도장마감 7 갈바 위 안티코 스터코 마감 8 갈바 위 안티코 스터코 마감 / 10mm 메지 9 갈바 위 흰색 쉬트 마감 / 선반 : 파이프 보강, 도장마감 10 냉장고 11 수전 : 자바라 12 기성 가구(흰색) 13 열연철판 취부 14 안티코 스터코(검정)마감 15 복도 16 열연철판 취부 17 모기장 취부 / 창호 단열재 내장 18 프레암:갈바 위 검정 도장 마감 / 두께 최소화 19 투명 유리 위 컬러 쉬트 + 프로스트 쉬트 취부 / 프로젝트 창(오도시)/모기장 취부/창호 단열재 내장 20 석고보드 위 V.P도장 마감 / 내부방음재(스티로폼) 21 천정메지 10X10mm 22 강화도어 : 손잡이 23 걸레받이 10X10mm : 유성도장 마감 24 계단 전면 : 갈바 위 도장 25 유리 홈파기/투명 유리 위 프로스트 쉬트 취부 26 강화도어 : 손잡이, 하부 시건장치 27 10X10mm sus(거울)유리 홈

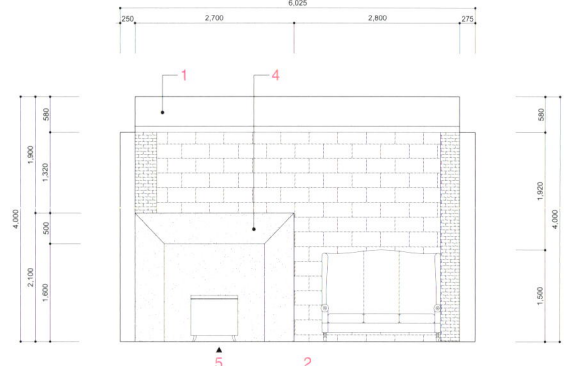 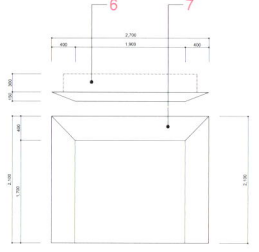 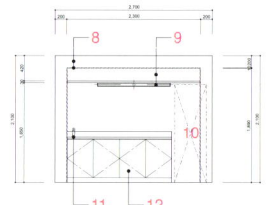

3층 대회의실 입면 L / 3rd floor conference room elevation L

3층 대회의실 입면 M / 3rd floor conference room elevation M

3층 대회의실 입면 N / 3rd floor conference room elevation N

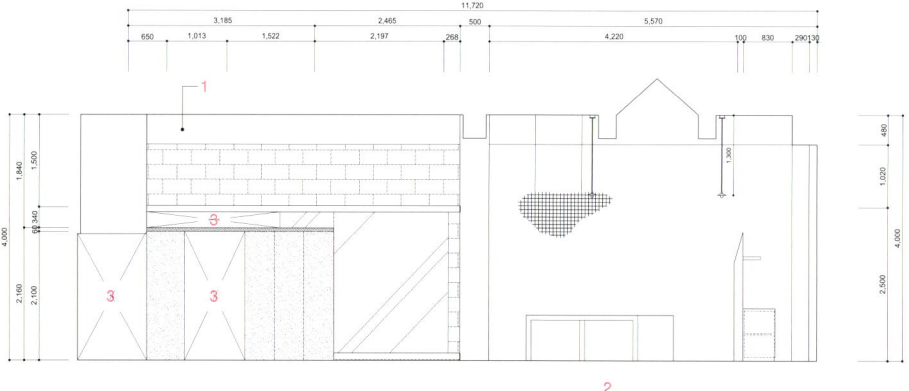

3층 대회의실 입면 O / 3rd floor conference room elevation O

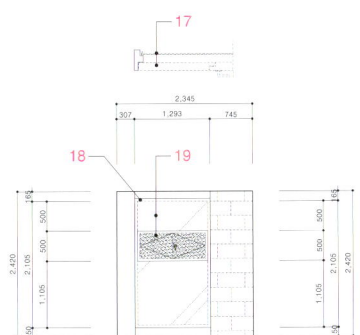

사장실 입면 Q / president room elevation Q

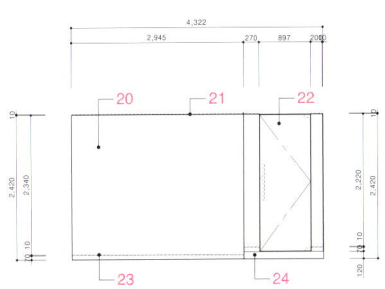

사장실 입면 R / president room elevation R

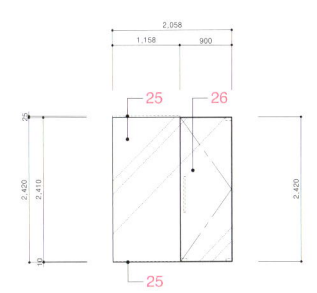

사장실 입면 S / president room elevation S

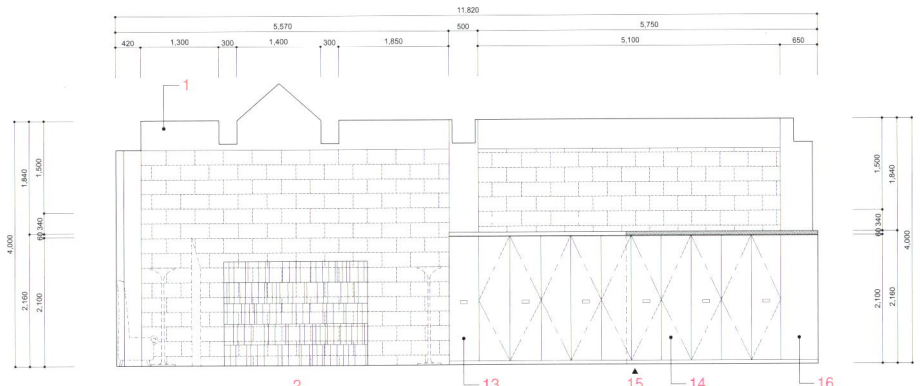

3층 대회의실 입면 P / 3rd floor conference room elevation P

KTF GALLERY-THE ORANGE

B613 디자인팀 | 정기태 B613 Design Team | Jung Ki Tae

KTF의 문화소통공간인 KTF Gallery - The orange는 1층의 KTF 이동통신 체험공간과 2-3층의 비영리 문화공간으로 이루어져 있다. 생명력과 창의력을 갖춘 예술가들에게 무료로 작품발표 공간으로 제공되면 공모형식을 통해 전시 작가를 선정할 계획이다. 전체 디자인 컨셉은 THE ORANGE로 창조와 감성의 메타포가 가득한 공간이다. 이는 새롭게 창조된 미지의 오렌지 나무를 대면하면서 호기심과 상상력으로 충만해지며 이 호기심 충만한 태동으로 첫 계단을 디디는 순간, 가슴 속에 따뜻한 감성으로 충만한 THE ORANGE 한 그루를 심게 된다. 이런 오렌지 나무의 성장 높이만큼 살포시 따라 2층과 3층에 도달했을 때, 기존과 다른 일탈적인 갤러리 공간에 스며들어 그 속에서 작품과 더불어 틀 속에 갇혀 있던 존재로서의 내가 아닌 창조적인 인간으로서의 나를 보게 된다.

THE ORANGE는 서로 상충되는 것이 모여, 조화 속에서 개성을 추구하는 진정한 상생의 공간으로 콘크리트 바닥과 벽돌과 구들장 등의 소박하고 고전적인 소재가 고급스러운 흑경과 절제된 백색의 곡선볼륨을 가진 미래적 형태인 벽과 어우러져 소통과 공존의 오묘한 묘미를 제공한다. 인간의 감성이 수분이 되어 자라 오르는 오렌지 나무는 다시 열매를 맺음으로써 삶의 열정을 선사함에 더불어 향기로운 공간인 THE ORANGE는 진정한 안식을 누릴 수 있는 감성적 향기를 제공할 것 이다.

THE ORANGE, the overall design concept, is a space replete with metaphors of creation and feelings. This is to fuel curiosity and imagination when we face the unknown orange tree that has been newly created and when we set foot on the first step of the stair, we will plant an orange tree filled with warm sentiment in our hearts. When we follow the growth of the orange tree to reach 2nd and 3rd floors, we slip into a gallery space, a getaway from existing spaces, and see ourselves not as people who are trapped in a certain mold but as creative human beings.

THE ORANGE is a genuine co-existential space that is different. The conflicting spaces are to get together to pursue their own individuality in harmony and provide exquisite charm of communication and coexistence. Simple and classic materials such as concrete floor and floor-slab are futuristic and restrained while the curved wall has a luxurious black mirror. As an orange tree that grows in human sentiment as its water gives the passion of life with its fruits, the fragrant space, THE ORANGE will provide a scent of sentiment to relish real rest and relaxation.

위치 서울시 중구 명동 2가 51-28 2-3F
면적 108㎡
설계 B613 DESIGNTEAM/정기태
설계팀 B613 DESIGNTEAM/김혜미, 최한철, 심선영
시공 동신토탈시스템/김평규
마감 바닥-투명 에포코트 벽체-V.P 도장. 조적 벽돌 천장-V.P 도장
사진 B613디자인팀 제공

Location 2-3F 51-28 Myeong-dong 2-ga, Jung-gu, Seoul, Korea
Building Area 108㎡
Design B613 DESIGNTEAM /Jung Ki Tae
Design team B613 DESIGNTEAM/Kim He Mi, Choe Han Cheol, Shim Sun Young
Construction DONGSIN TOTAL SYSTEM /Kim Pyeong Gyo
Finishing floor-clear epocoat wall-V.P. Brick laying ceiling- V.P
Photos offer B613 Design Team

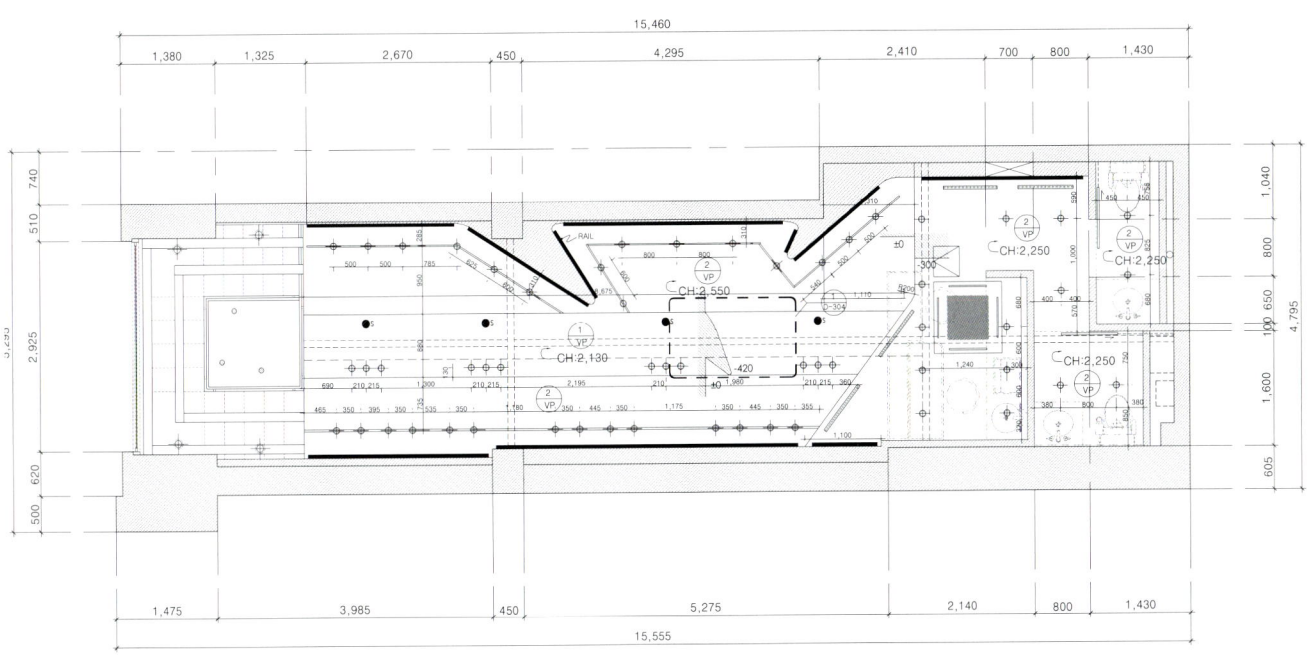

2층 천정도 / 2nd ceiling plan

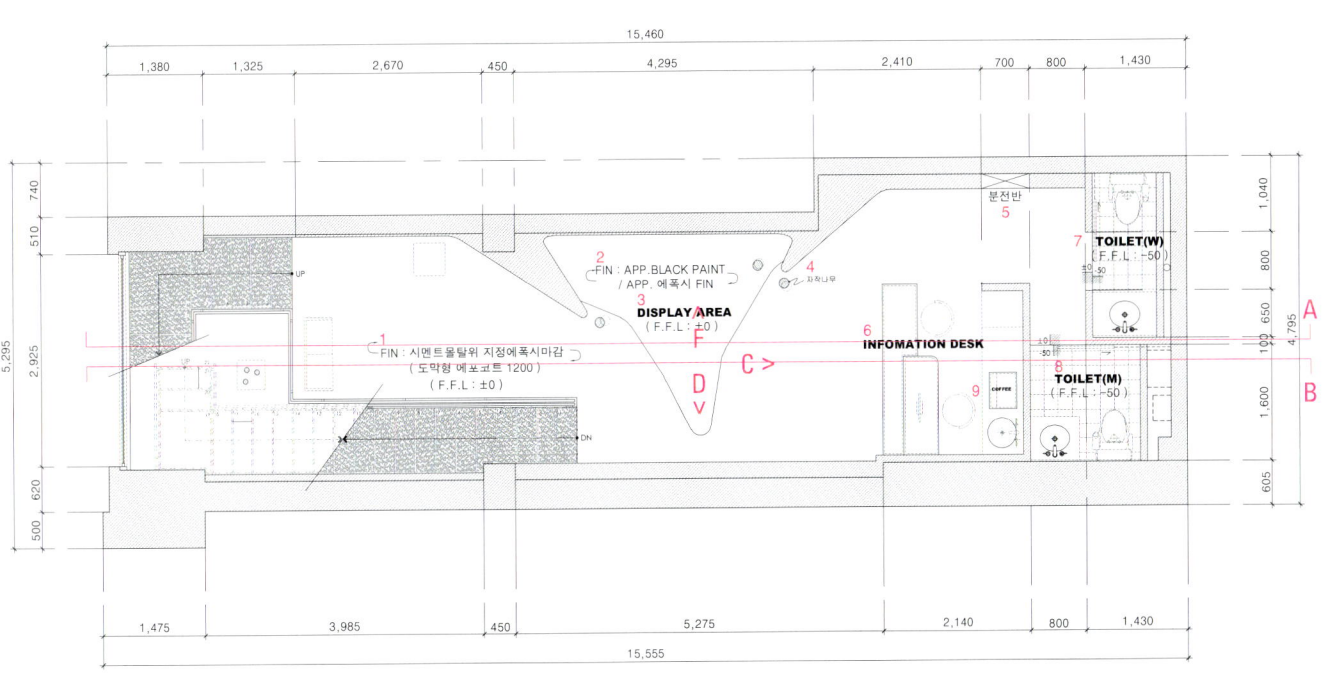

2층 평면도 / 2nd floor plan

1 Fin. : App. epoxy Fin. on cement mortar(surface type epocoat 1,200)(F.F.L : ±0) 2 Fin. : App. Black Paint, App. epoxy Fin. 3 Display Area(F.F.L : ±0) 4 white birch 5 cabinet panel 6 Information Desk 7 Toilet((F.F.L : -50) 8 Toilet(F.F.L : -50)

1 마감 : 시멘트몰탈위 지정에폭시마감(도막형 에포코트 1,200)(F.F.L : ±0) 2 마감 : 지정 검정 페인트, 지정 에폭시 마감 3 디스플레이 공간(F.F.L : ±0) 4 자작나무 5 분전반 6 인포메이션 데스크 7 화장실((F.F.L : -50) 8 화장실(F.F.L : -50)

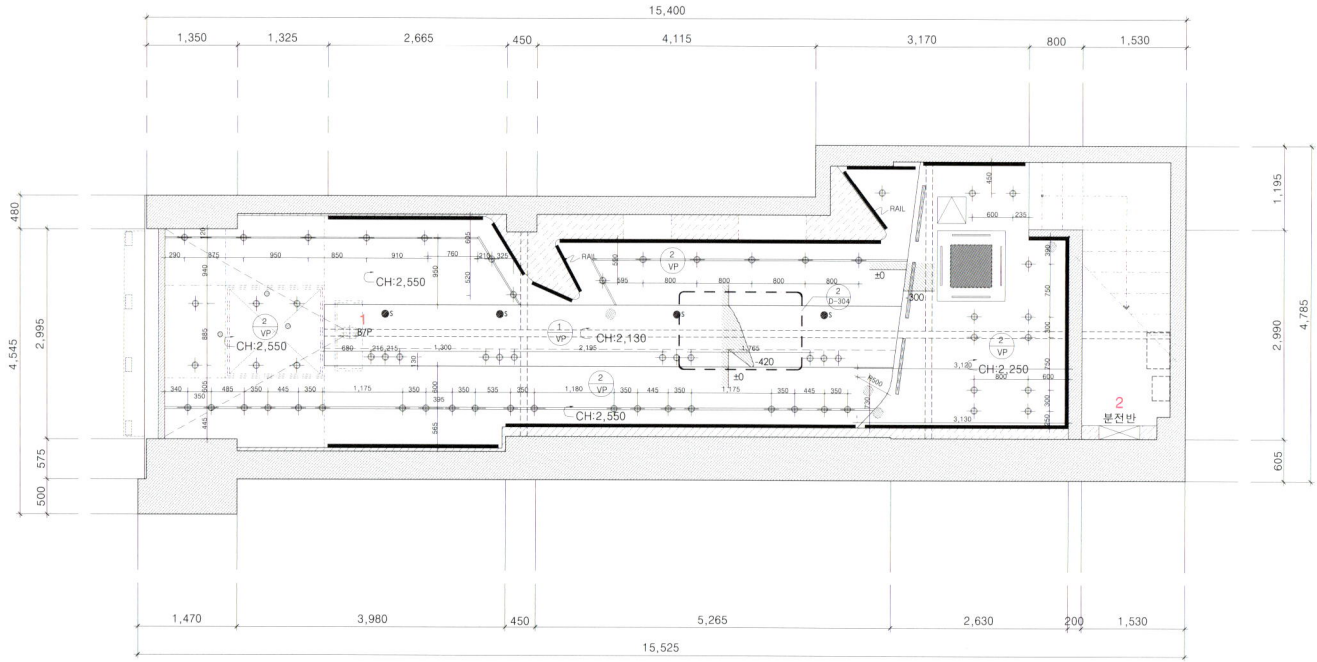

3층 천정도 / 3rd ceiling plan

1 B/P 2 cabinet panel 3 Display Area A 4 white birch 5 Fin. : App. epoxy Fin. on cement mortar(surface type epocoat 1,200)(F.F.L : ±0) 6 Display Area B 7 ㅁ-30X30 S'STL Pipe / T1.6mm Galva Plate 8 ㅁ-30X30 S'STL Pipe 9 T1.6mm Galva Plate 10 Neon Lighting

1 B/P 2 분전반 3 디스플레이 공간 A 4 자작나무 5 마감 : 시멘트 몰탈 위 지정 에폭시 마감(도막형 에포코트 1,200)(F.F.L : ±0) 6 디스플레이 공간 B 7 ㅁ-30X30 S'STL 파이프 / T1.6mm 갈바판 8 ㅁ-30X30 S'STL 파이프 9 T1.6mm 갈바판 10 네온 조명

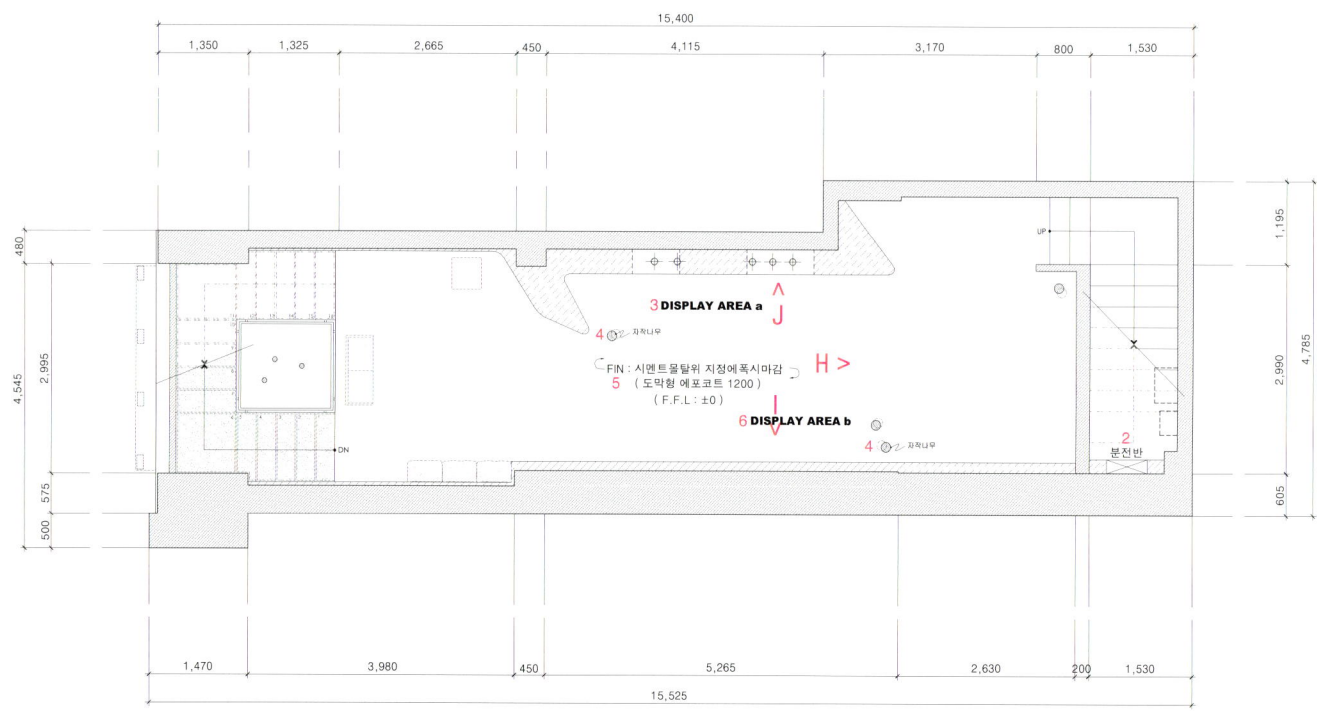

3층 평면도 / 3rd floor plan

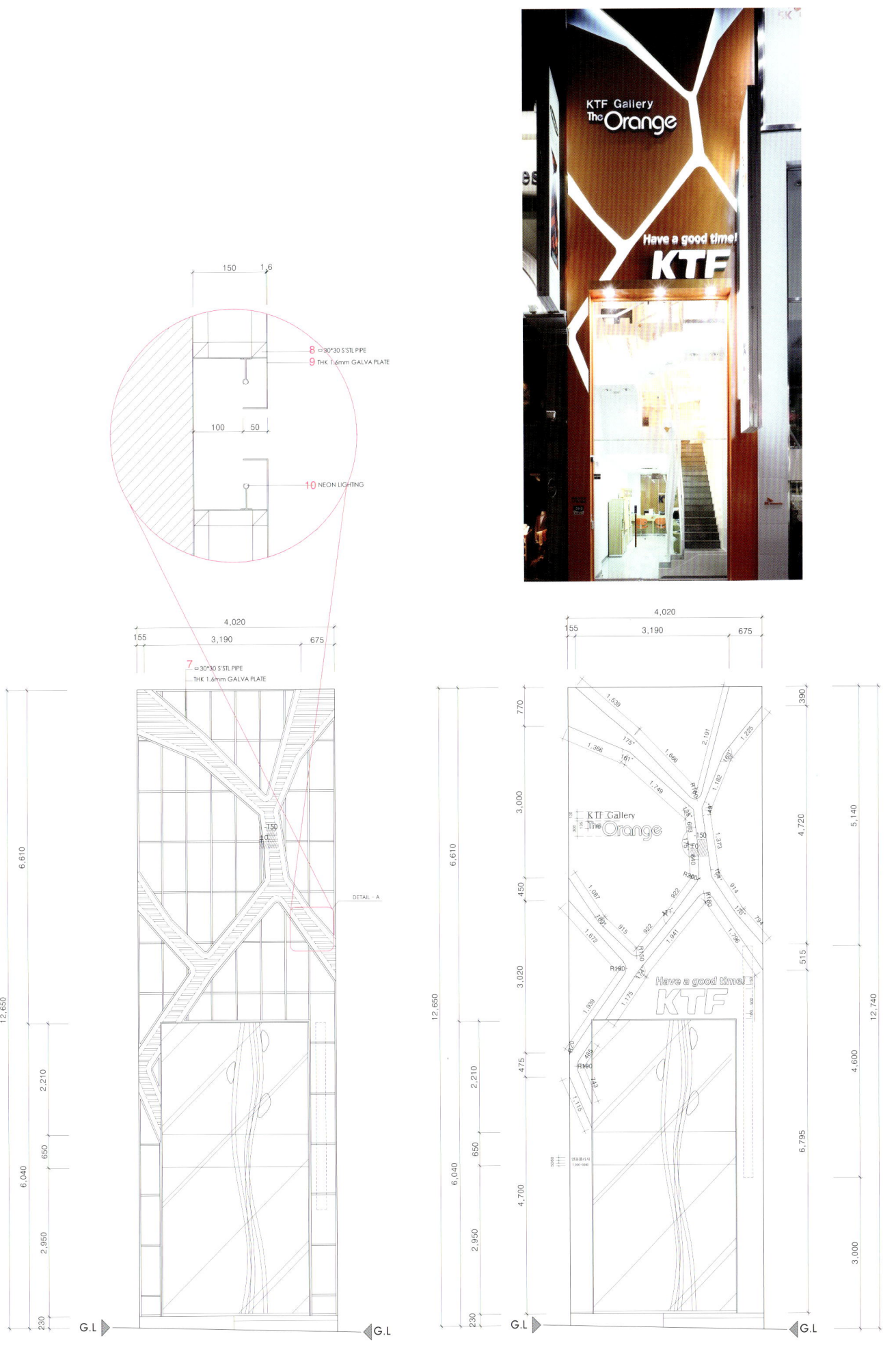

파사드 입면 / facade elevation

KTF GALLERY-THE ORANGE | KTF 갤러리-오렌지

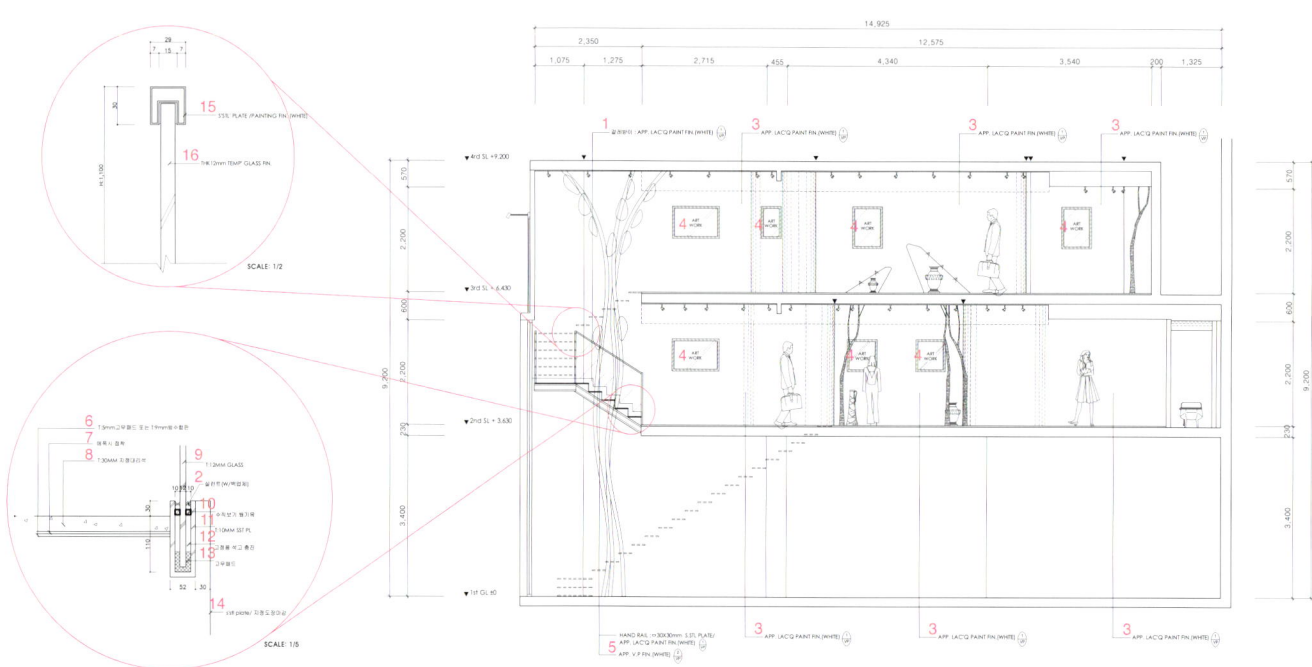

계단 상세 / stair detail

단면 A / cross section A

1 baseboard : App. Lacq' Paint Fin.(White) 2 Sealant(W / backup material) 3 App. Lacq' Paint Fin.(White) 4 Art Work 5 Hand Rail : □-30X30mm S'STL Plate, App. Lacq' Paint Fin.(White) / App. V.P Fin.(White) 6 T5mm rubber pad or T9mm waterproof plywood 7 epoxy bonding 8 T30mm appointed marble 9 T12mm Glass 10 perpendicular 쐐기목 11 Thk10mm SST PL 12 filling gypsum fixing 13 rubber pad 14 S'STL Plate, appointed painting finish 15 S'STL Plate, Painting Fin.(White) 16 T12mm Temp' Glass Fin. 17 T5mm Black Mirror Fin. 18 App. Cemeent Block Fin.(250X540) 19 Hand Rail : □-30X30mm S'STL Plate, App. Lacq' Paint Fin.(White) / App. V.P Fin.(White) 20 Sliding 21 App. basalt Fin.

1 걸레받이 : 지정 락커 페인트 마감(흰색) 2 실런트(W / 백업재) 3 지정 락커 페인트 마감(흰색) 4 아트작업 5 핸드레일 : □-30X30mm S'STL 판, 지정 락커 페인트 마감(흰색)/ 지정 도장 마감(White) 6 T5mm 고무 패드 또는 T9mm 방수합판 7 에폭시 접착 8 T30mm 지정대리석 9 T12mm 유리 10 수직보기 쐐기목 11 T10mm SST PL 12 고정용 석고 충진 13 고무패드 14 S'STL 판, 지정 도장 마감 15 S'STL 판, 도장 마감(흰색) 16 T12mm 강화 유리마감 17 T5mm 흑경 마감 18 지정 시멘트 블럭 마감(250X540) 19 핸드레일 : □-30X30mm S'STL 판, 지정 락커 페인트 마감(흰색) / 지정 도장 마감(흰색) 20 슬라이딩 21 지정현무암 마감

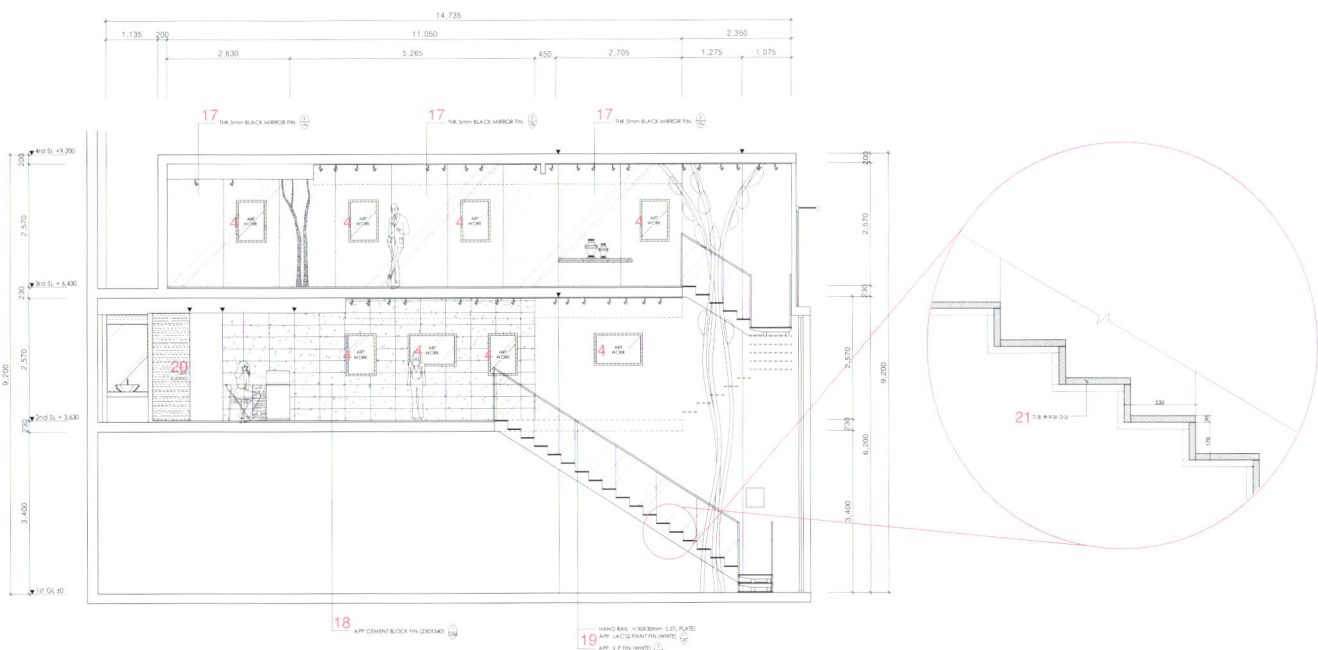

단면 B / cross section B 계단 상세 / stair detail

1 Speaker attaching 2 App. V.P Fin.(White) 3 App. epoxy Fin. on App.cement mortar 4 App. Urethane Paint Fin.(White) 5 Sliding 6 Door : App. Wood Film Fin. 7 App. Cement Block Fin.(250X540) 8 Hand Rail : ▫-30X30mm S'STL Plate, App. Lacq' Paint Fin.(White) 9 Object : App. FRP Fin. 10 Art Work 11 App. Lacq' Paint Fin.(White) 12 App. epoxy Fin. on cement mortar(surface type epocoat 1,200) 13 baseboard : App. Lacq' paint Fin.(White) 14 Open to Toilet 15 Object : App. FRP Fin.

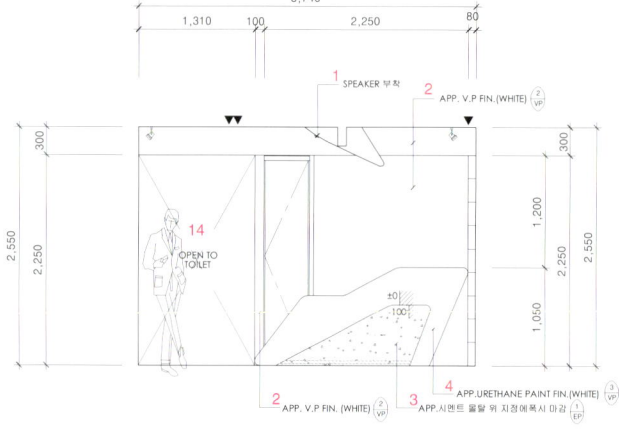

2층 입면 C / 2nd floor elevation C

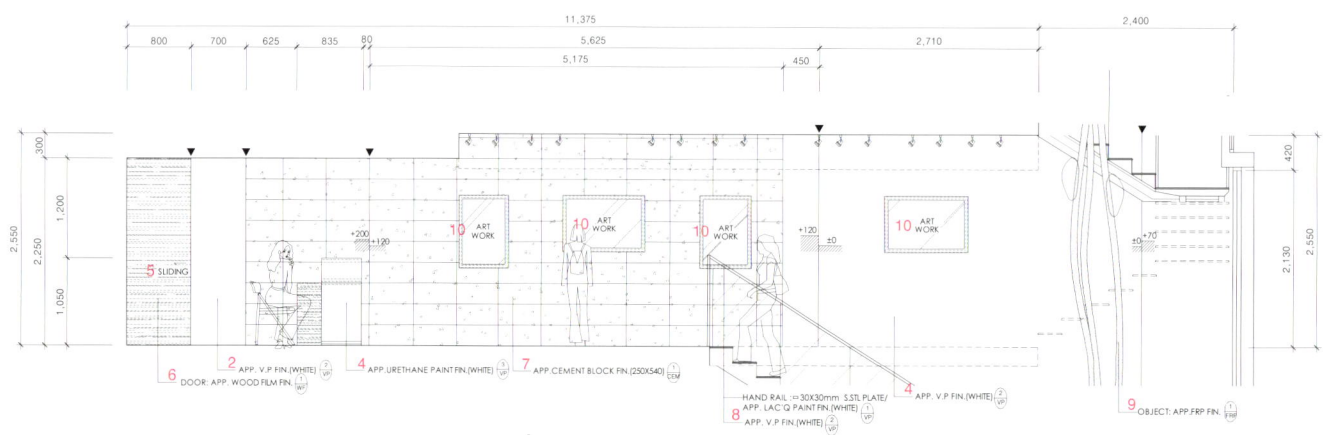

2층 입면 D / 2nd floor elevation D

1 스피커 부착 2 지정 도장 마감(흰색) 3 지정 시멘트 몰탈 위 지정 에폭시 마감 4 지정 우레탄 페인트 마감(흰색) 5 슬라이딩 6 문 : 지정 우드 필름 마감 7 지정 시멘트 블록 마감(250X540) 8 핸드레일 : ㅁ-30X30mm S'STL 판, 지정 도장 마감(흰색) 9 물체 : 지정 FRP 마감 10 아트 작업 11 지정 도장 마감(흰색) 12 지정 시멘트 몰탈 위 지정에폭시 마감(도막형 에포코트 1,200) 13 걸레받이 : 지정 도장 마감(흰색) 14 화장실 열린공간 15 물체 : 지정 FRP 마감

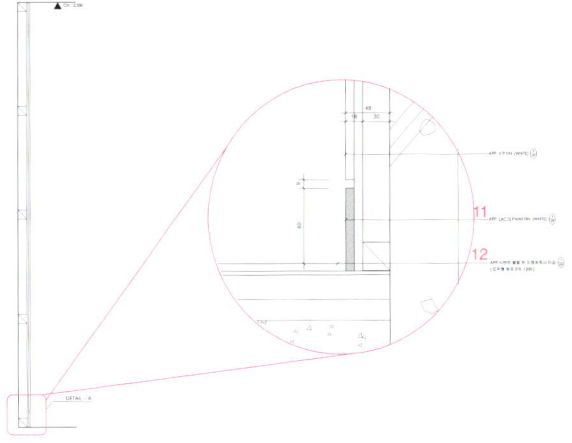

벽 상세 / wall section

2층 입면 F / 2nd floor elevation F

1 Speaker attaching 2 App. Lacq' paint Fin.(White) 3 Thk5mm Clear Glass / App. Stone Fin. 4 Object : App. FRP Fin. 5 Handrail : □-30X30mm S'STL Plate, App. Lacq' Paint Fin.(White) 6 Thk12mm Temp' Glass 7 T80 S'STL Pipe, App. Lacq' Paint Fin.(White) 8 App. FRP Fin, Color Fin.(Orange) 9 volt hole 10 App. FRP Fin.

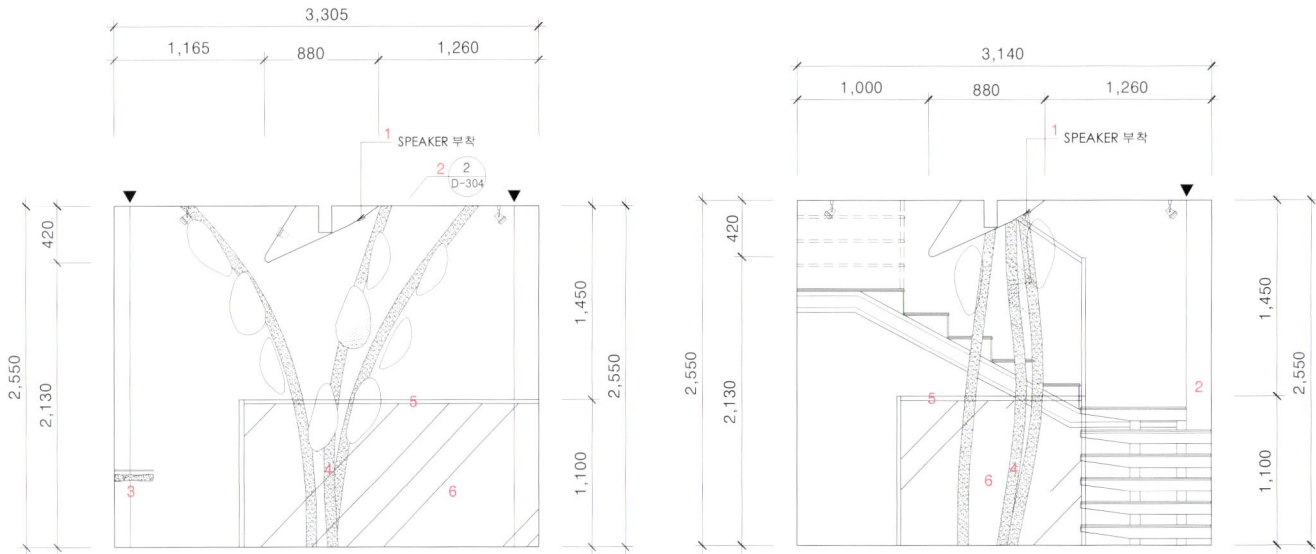

3층 오브제 입면 / 3rd object elevation

2층 오브제 입면 / 2nd object elevation

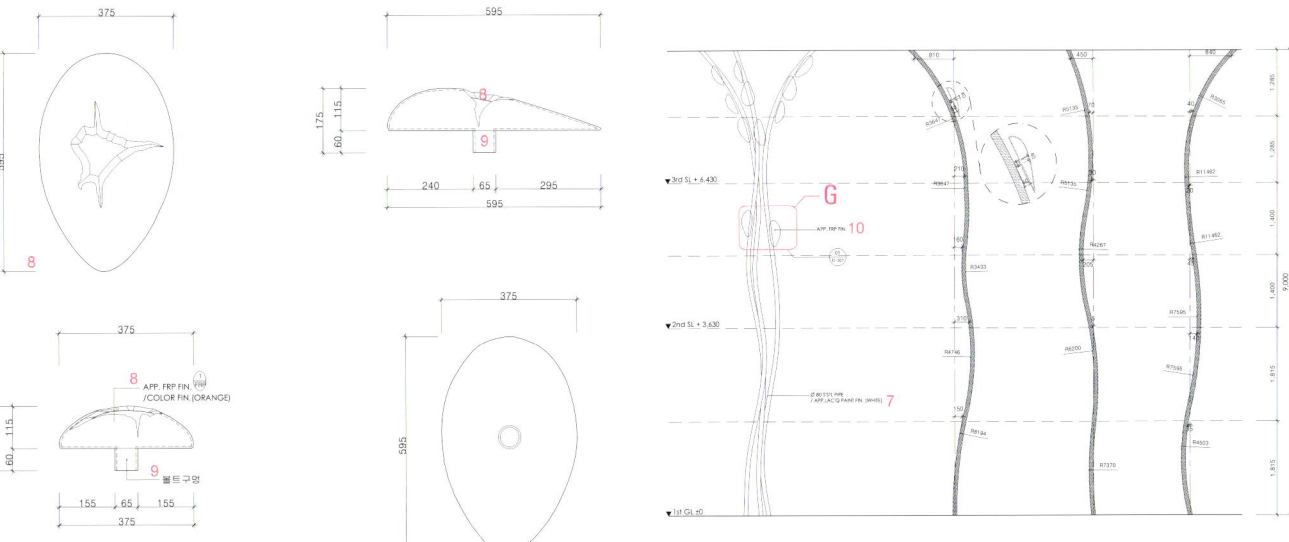

1 스피커 부착 2 지정 락커 도장 마감(흰색) 3 T5mm 투명 유리/ 지정 돌 마감 4 물체 : 지정 FRP 마감 5 난간 : ㅁ-30X30mm S'STL 판, 지정 락커 도장 마감(흰색) 6 T2mm 강화 유리 7 T80 S'STL 파이프, 지정 락커 도장 마감(흰색) 8 지정 FRP 마감, 컬러 마감(오렌지) 9 볼트구멍 10 지정 FRP 마감

오브제 상세 G / object detail G

오브제 입면 상세 / object elevation section

AN ANNEX TO THE ASIA PUBLICATION CULTURE AND INFORMATION CENTER

전 어소시에이트 | 전시형 Jeon Asscociates I Jeon Shi Hyoung

광활한 공간에 빛을 가진 벽을 세움으로써 기존의 공간을 최대한 살리면서 조화될수 있는 방향에 초점을 맞추었다. 지지향이라는 게스트하우스의 자연스러우면서 은은한 이미지를 표현하기 위해 조명의 색상과 빛이 퍼지는 느낌을 적절하게 표현하는데에 신중을 기했다. 빛과 화이트 Corian의 담백하고 은은한 느낌은 시각적으로 임팩트있는 새로움을 가지면서도 기존 공간과 조화되어 생동감과 역동성을 더해주는 역할을 한다. 형태적인 부분에 있어서도 박스형태의 벽이 아닌 사선과 각도를 이용해 매스를 구성하여 공간 안에서 벽이 아닌 오브제와 같은 역할을 하고 있다. 또한 동일한 벽이 연속적으로 펼쳐지기 보다는 주변의 자연환경을 부분적으로 수용하고 내부 공간의 이미지를 구성할 수 있게끔 비례감을 고려하여 공간 안에 wall을 세워 기존의 마감재와 자연스럽게 연결되도록 하였다. 이런 형태를 가진 빛 덩어리를 주축으로 기능을 하는 desk나 bar와 같은 부분은 정직한 형태의 덩어리로 무게감과 안정감을 표현하였다. 바닥 러그를 통해 공간의 연결성을 유지하되 그 위에 각각의 소파를 배치하여 독립적인 zone을 형성하면서도 공간의 분할이 최소한이 되도록 배치하였다.

By erecting a lighting wall in a vast area, it seeks to focus on creating a harmony with the existing space without reducing it to the background. The simple and subtle feeling emanating from the light and white corian delivers a visually novo impact while playing a role in adding vivacity and a dynamic sense in harmony with the existing space. In terms of shape, it escapes from a box-form wall but composes a mass with oblique lines and angles to assign an objet-like role to the wall in the space. Also, rather than resorting to an identical and consecutive wall, it partly embraces the surrounding natural environment and seamlessly connects it with the eliminating finishing materials by erecting a wall in a place in consideration of its proportion to create an image of the interior space. Parts whose functions center on such form of light mass such as a desk and bar assume a straight form to express a sense of weight and security. The floor rug seeks to keep a sense of continuity in the space while arranging each sofa on it to create independent zones while minimizing the division of space.

위치 경기도 파주시 교하읍 문발리 파주출판도시 524-3
면적 1,300㎡
설계, 시공 전 어소시에이트
디자인팀 송미선, 안현진
완공 2007. 5
건축주 출판도시문화재단
마감 벽-코리안, 바리솔 바닥-마루에 가죽러그 마감
사진 전 어소시에이트 제공

Location 524-3 Pajuchulpandosi, Munbal-ri, Gyoha-eup, Paju-si, Gyeonggi-do, KOREA
Floor area 1,300㎡
Design, Construction Jeon Asscociates
Design team Song Mi Sun, An Hyun Jin
Completion 2007. 5
Client Bookcity Culture Foundation
Finishing Wall-Corian, Barrisol Floor-Skin rug on wooden floor fin
Photos offer Jeon Associates

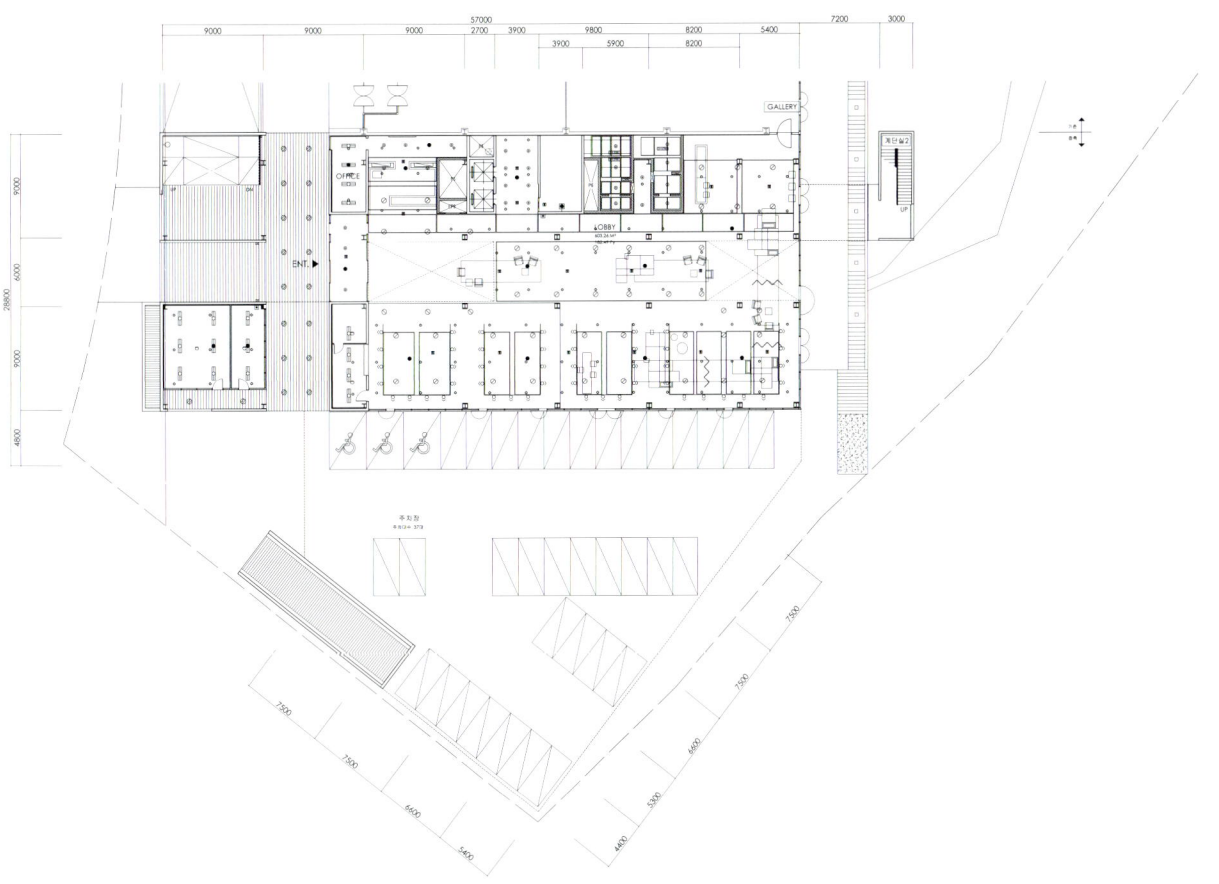

로비 천정도 / lobby ceiling plan

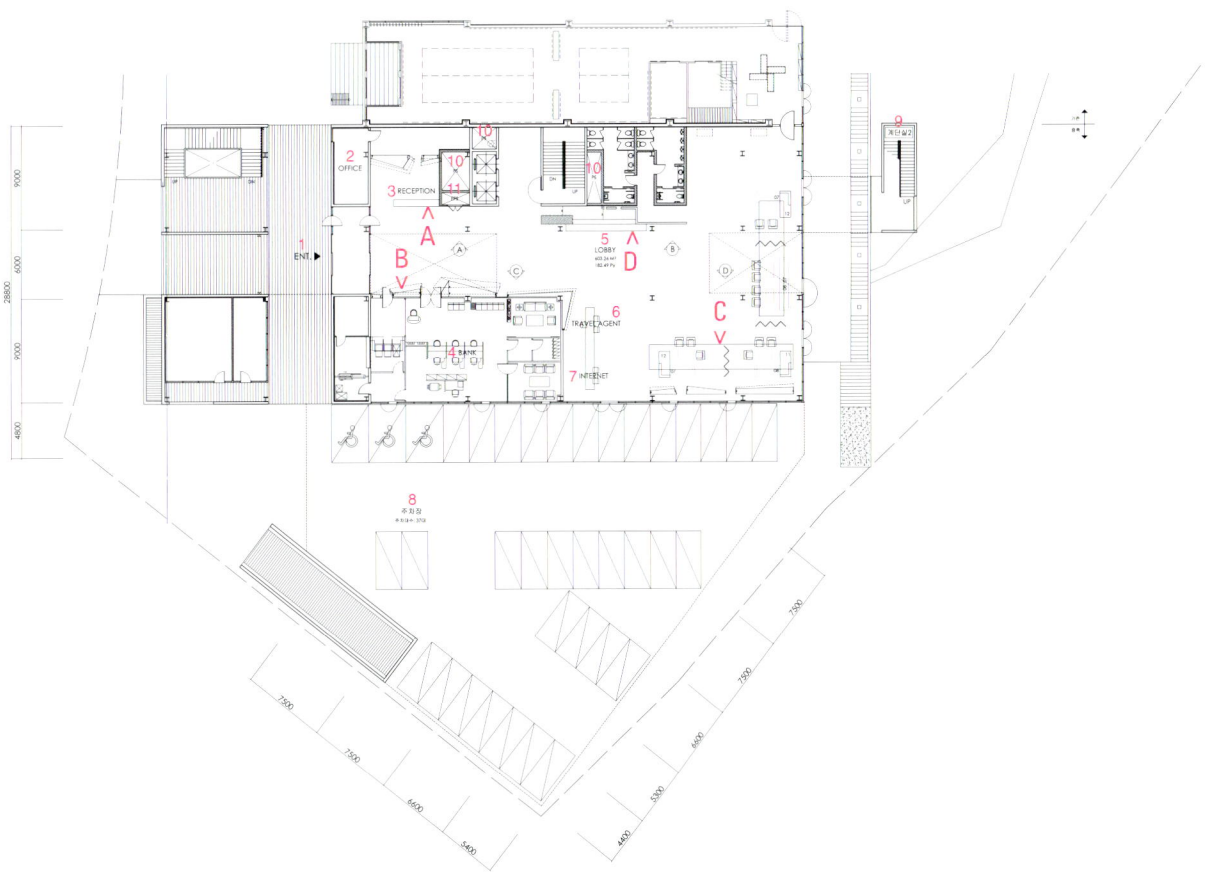

로비 평면도 / lobby floor plan

1 ENT.　2 Office　3 Reception　4 Bank　5 Lobby　6 Travel Agent　7 internet zone　8 car park　9 staircase　10 PS　11 EPS　12 Gallery

1 입구　2 사무공간　3 리셉션　4 은행　5 로비　6 여행사　7 인터넷존　8 주차장　9 계단실　10 PS　11 EPS　12 갤러리

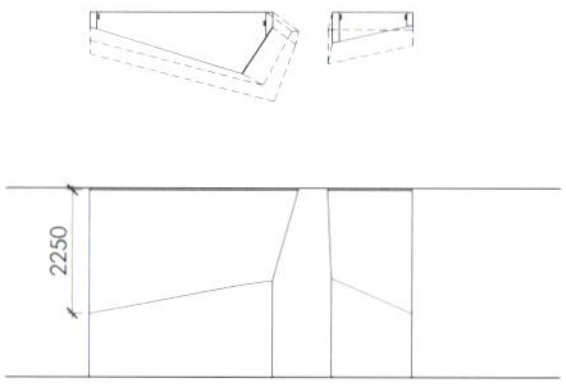

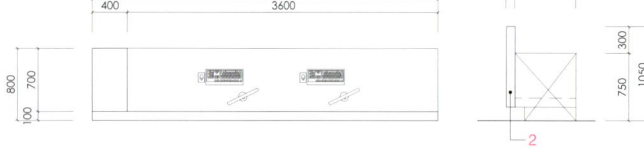

리셉션 데스크 정면 & 측면 / reception desk front view & side view

리셉션 데스크 평면 & 단면 / reception desk front plan & section

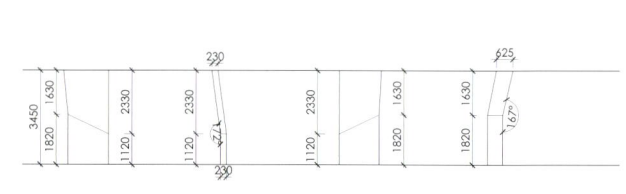

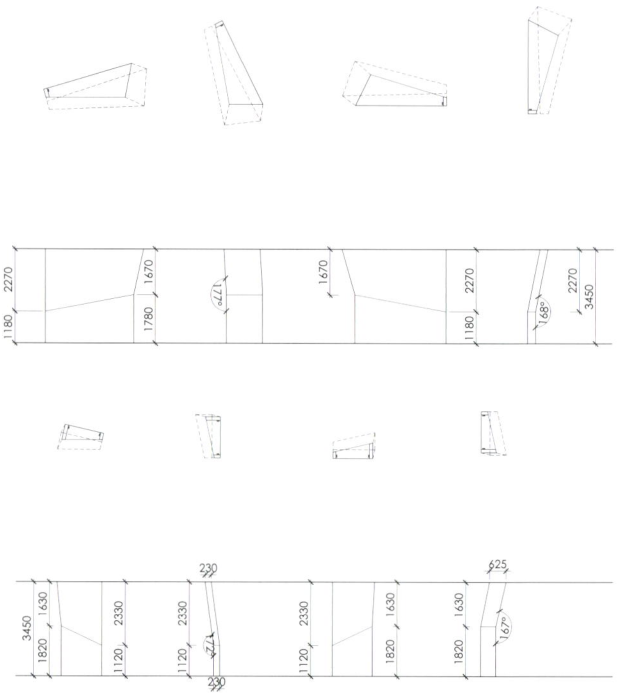

리셉션 바리솔 기둥 상세 / reception barrisol pillar section

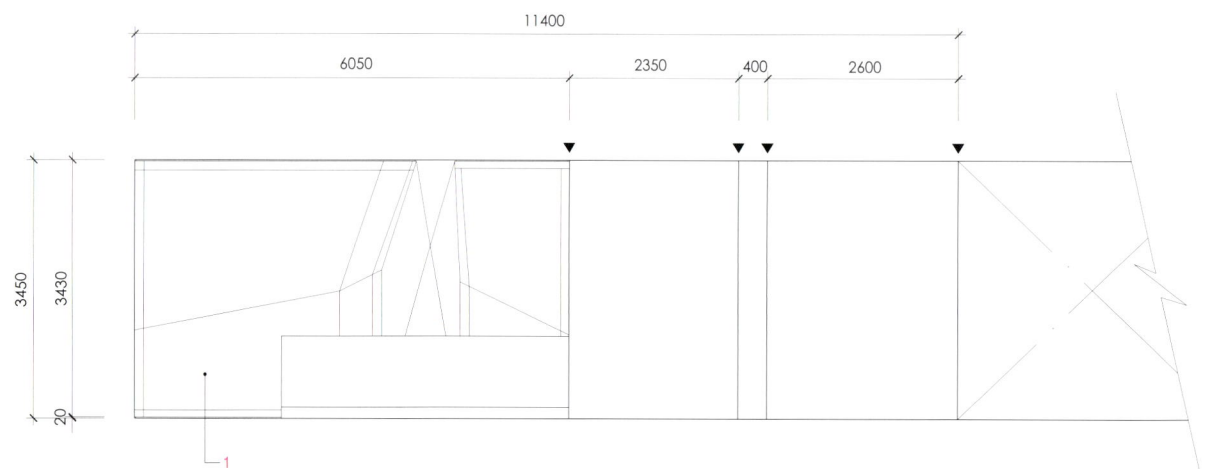

리셉션 입면 A / reception elevation A

1 Wall 2 Corian 3 Susmrror 4 Lighting

1 벽 2 코리안 3 Susmrror 4 조명

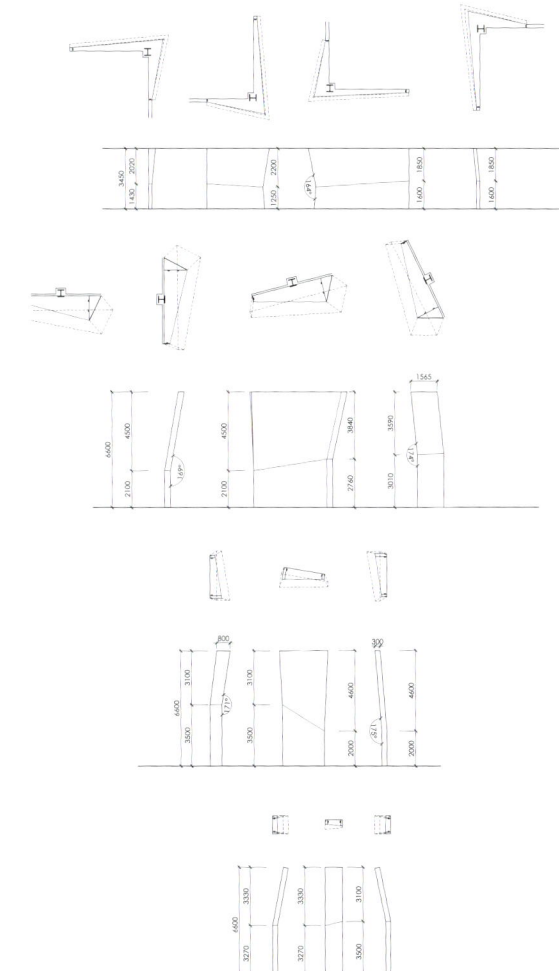

은행 입구 입면 B / bank entrance elevation B

은행 입구 바리솔 기둥 상세 / bank entrance barrisol pillar section

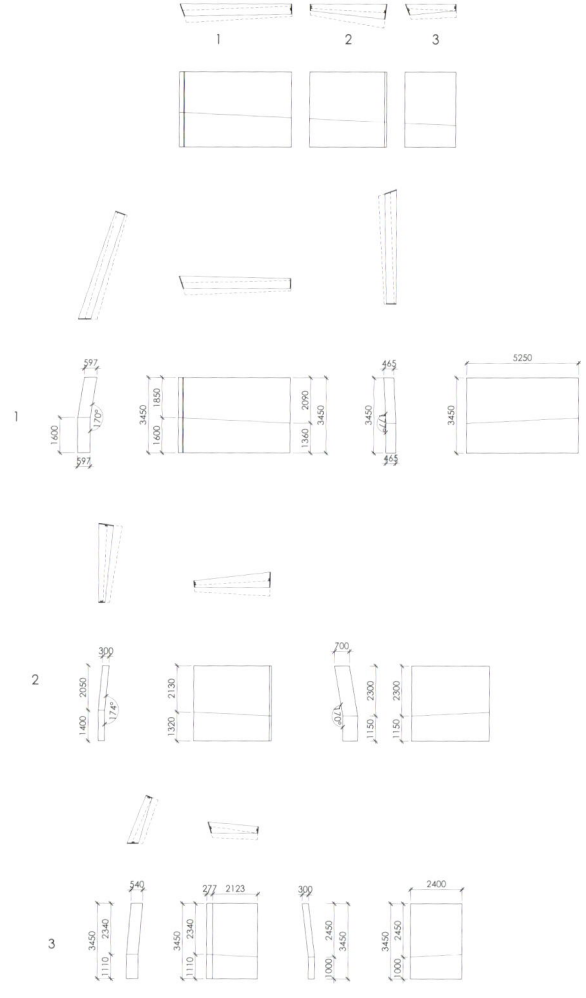

창측 바리솔 기둥 상세 / window part barrisol pillar section

창측 입면 C / window part elevation C

1 Lighting 2 Corian 3 Susmrror

1 조명 2 코리안 3 Susmrror

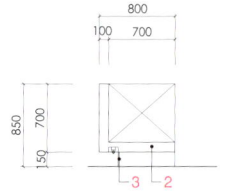

여행사 카운터 평면 & 측면 / travel agency counter plan & side view

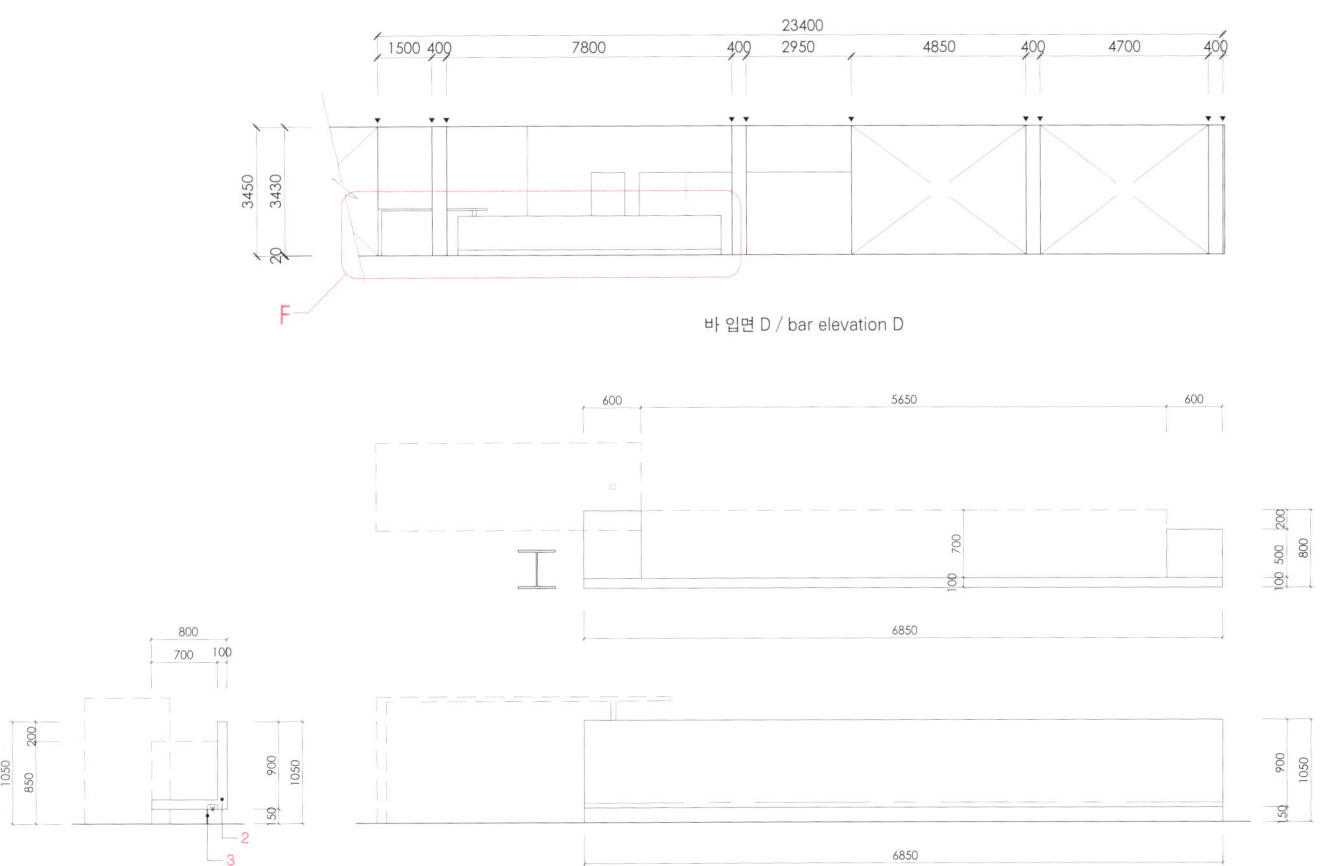

바 입면 D / bar elevation D

바 평면 & 정면 & 단면 / bar plan & front view & section

LITTLE BEAR

프랜즈 디자인 | 양진영   Friends Design | Yang Jin Young

키즈까페 리틀베어는 온 가족이 함께 즐기고 쉴 수 있는 복합 문화 공간으로 엄마를 위한 공간과 아이들을 위한 공간 두 가지로 분류됨과 동시에 시각적 연결을 통해 서로 연결시켜준다. 기존의 어린이 놀이공간에서 한층 업그레이드되어 기능적, 디자인적, 프로그램적 요소 등으로 차별화 된 공간을 조성하여 좀 더 아이들과 가까워 질 수 있는 공간을 디자인 하였다. 여섯 번째 리틀베어 프로젝트 동백점은 기존 리틀베어의 색채감과 함께 부드러운 파스텔톤의 다채로운 색채와 어린이 놀이 공간에 다양한 프로그램을 접목하여 아이의 시선에 맞춘 공간을 표현 하고자 하였다. 다채로운 컬러의 어린이 놀이기구와, 파티룸, 놀이방, 벽화로 마감한 아이들이 모습을 촬영해 그 순간을 기억할 수 있게 한 포토존은 아이들을 위한 공간이고, 홀과 테라스에 마련된 카페테리아는 엄마들을 위한 공간으로 아이들이 뛰노는 모습을 한 눈에 볼 수 있으면서도 자신의 시간을 가질 수 있는 공간으로 마련되었다. 공간 속에 나무 오브제를 두어 외부에서 내부로 자연을 가져옴으로써 보는 이들로 하여금 좀 더 쾌적하고 편안함으로 다가올 수 있도록 하였다. 전체적으로 밝고 경쾌함 느낄 수 있도록 하면서 하나의 공간으로 표현이 될수 있도록 전체적으로 다채로운 색감이 분포 되도록 하였다. 공간별로 다양한 시트작업을 통해 재미있는 요소 뿐 아니라 어린이들의 상상과 호기심을 자극하여 창의력 발달에 도움이 되도록 하여 아이들이 한 공간에 머물지 않고 다양하게 돌아다닐 수 있는 공간으로 계획 하였다. 아이들이 마음껏 뛰어 놀 수 있는 공간이면서 엄마들에게도 편안한 휴식공간을 제공하는 공간을 제안하는 리틀베어는 '따로 또 같이'라는 공간의 의미를 표현하고자 했다.

Little Bear, kids' cafe where is divided into two zones for kids and for moms respectively connects moms and kids through the visual integrity of the spaces. The cafe which was upgraded with better functional, design, and program elements was designed to get one step closer to kids. The sixth Little Bear Project for Dongbaek branch was designed to integrate soft pastel tone palette that represents the brand Little Bear with various programs provided to kids. The photo zone, colorfully decorated with some themes such as rides, party room, playing room was made to show pleasant playing time of the kids to other kids and moms while the cafeterias placed in the hall and terrace were built for moms to enjoy their own coffee time, looking over the kids playing in the cafe from time to time. By placing an wooden object, users can feel as if the nature is moved into the space. In general, the cafe was designed as a bright, vivid and cheerful space in various colors. And such intriguing decoration sheets were added to every space to captivate kids's eyes, so that kids will not be easily bored with plain color or design of a room. Little bear where kids can play freely and moms can enjoy a comfortable and relaxing time intends to provide an 'independent but still dependent' space to both of kids and moms.

위치 경기도 용인시 기흥구 중동 851-4 동백역타워 9층
면적 288.18㎡ / 87.17PY(계단, 테라스일부분, ELEV제외)
설계기간 2007. 7. 1 ~ 7. 15
시공기간 2007. 7. 16 ~ 8. 17
설계 프랜즈디자인 / 양진영, 권일권
시공 프랜즈디자인 / 장영권, 이연봉, 김소희
마감 바닥- 데코타일, 카펫타일, 인조잔디, 고무매트 벽체-V.P도장, V.P 컬러도장, 시트마감, 소부도장 유리, 벽화 천정- V.P도장, 벽화
사진 프랜즈 디자인 제공

Location 851-4 Jung-dong Giheung-gu Yongin-si Gyeonggi-do, Korea
Area 288.18㎡
Design period 2007. 7.1 ~ 7.15
Design & Construction Friends Design
Construction period 2007. 7.16 ~ 8.17
Finishing Floor-Deco Tile, Carpet Tile, Artificial turf, mat Wall-Vinyl Paint, Sheet, Color Glass, Wall painting Ceiling-Vinyl Paint, Wall painting
Photos offer Friends Design

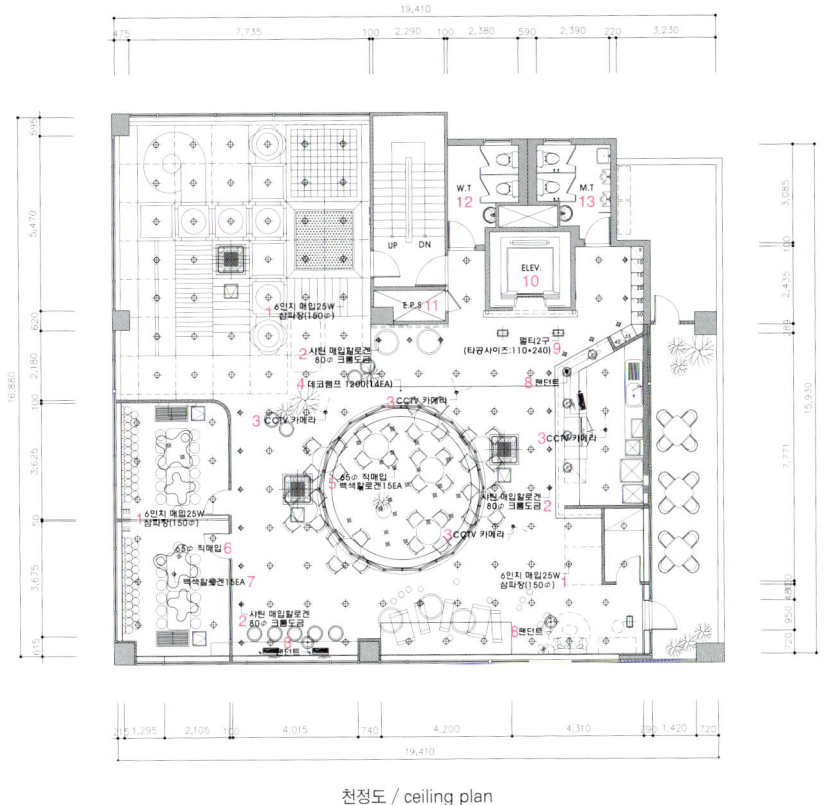

천정도 / ceiling plan

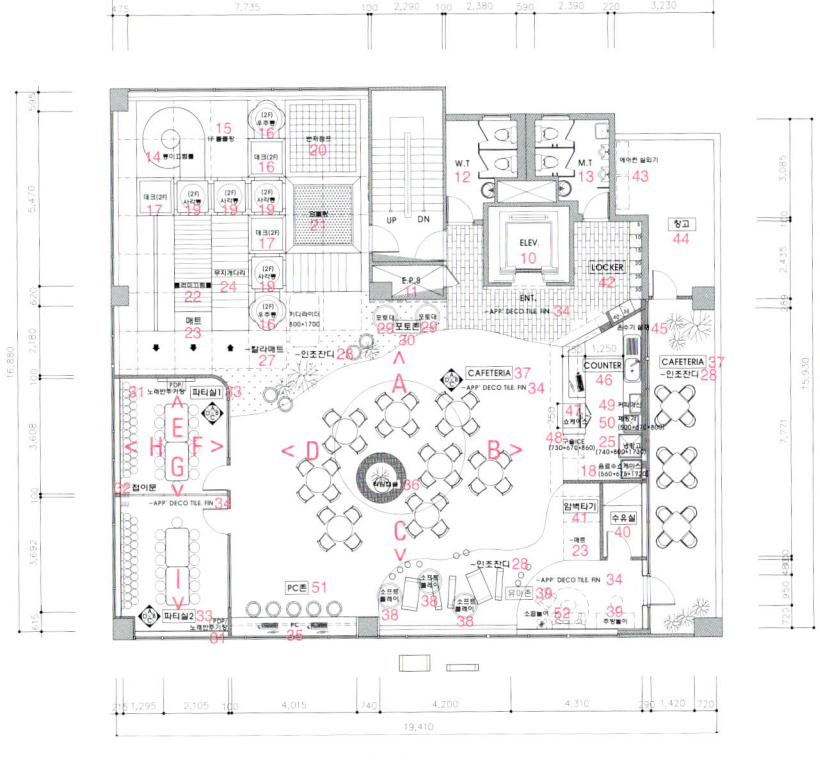

평면도 / floor plan

1 6 inch burying 25W three-cornered lamp(150) 2 satin burying halogen light 80 chrome plate 3 CCTV camera 4 deco lamp 1,200(13EA) 5 65 direct burying white halogen 15EA 6 65 direct burying 7 white halogen 15EA 8 pendant 9 multi 2places(Punch size : 110X240) 10 ELEV. 11 E.P.S 12 W.T 13 M.T 14 slide 15 1F ball pool 16 2F universe cylinder 17 2F deck 18 drinking water showcase (560X675X1,720) 19 2F square 20 bungee jump 21 dumbling 22 roller slide 23 mat 24 rainbow bridge 25 refrigerator(740X800X1,730) 26 Kiddy Rider(800X1,700) 27 color mat 28 artificial turf 29 photo shelf 30 photo zone 31 PDP, singing machine 32 Folding door 33 party room 34 App. Deco Tile Fin 35 PC 36 time capsule 37 Cafeteria 38 soft Bley 39 baby zone 40 nursing room 41 rock-climbing 42 Locker 43 air conditioner outdoor machine 44 storage 45 installing water heater 46 Counter 47 showcase 48 marble ICE(730X670X860) 49 coffee machine 50 ice machine (500X670X800) 51 PC aone 52 playing house

1 6인치 매입25W 삼파장(150) 2 샤틴 매입할로겐 80 크롬도금 3 CCTV 카메라 4 데코램프 1,200(13EA) 5 65 직매입 백색할로겐 15EA 6 65 직매입 7 백색할로겐 15EA 8 팬던트 9 멀티2구(타공사이즈 : 110X240) 10 ELEV. 11 E.P.S 12 W.T 13 M.T 14 통미끄럼틀 15 1F 볼플장 16 2F 우주통 17 2F 데크 18 음료수쇼케이스(560X675X1,720) 19 2F 사각통 20 번지점프 21 덤블링 22 롤러미끄럼틀 23 매트 24 무지개다리 25 냉장고(740X800X1,730) 26 키디라이더(800X1,700) 27 칼라매트 28 인조잔디 29 포토대 30 포토존 31 PDP, 노래반주기장 32 접이문 33 파티실 34 지정 데코타일 마감 35 PC 36 타임캡슐 37 카페테리아 38 소프트 블레이 39 유아존 40 수유실 41 암벽타기 42 락커 43 에어컨 실외기 44 창고 45 온수기 설치 46 카운터 47 쇼케이스 48 구슬ICE(730X670X860) 49 커피머신 50 제빙기(500X670X800) 51 PC존 52 소꿉놀이

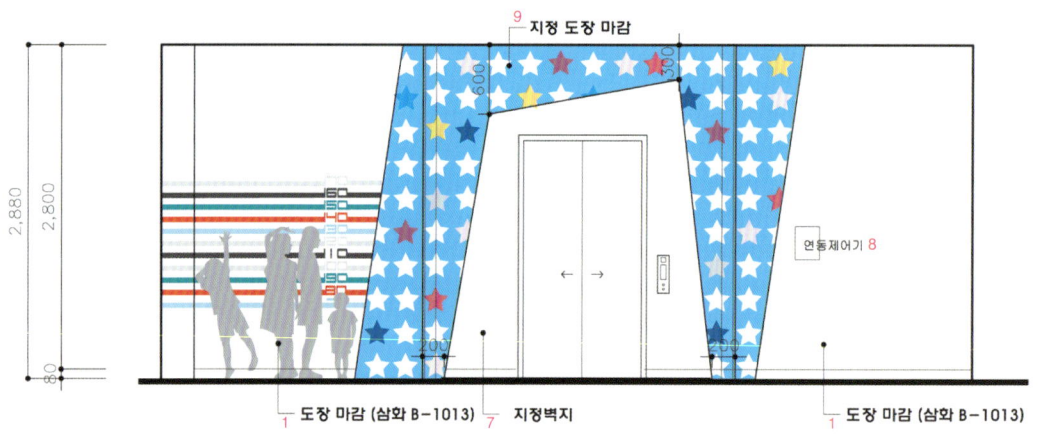

입구 그래픽 / entrance graphic

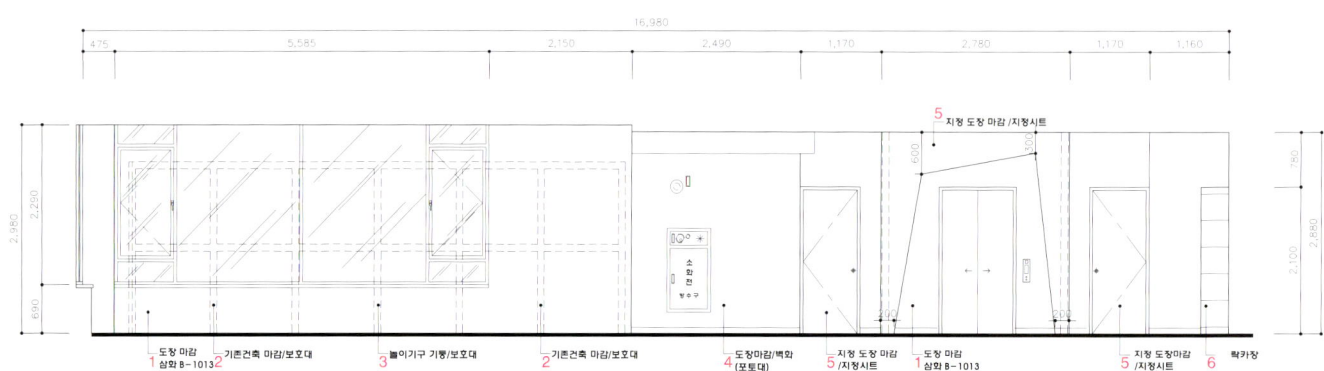

홀 입면도 A / hall elevation A

1 finishing by appointed lacquering samhwa B-1013 2 existing building finish, protector 3 rides pillar, protector 4 finishing by lacquering, wall painting 5 finishing by appointed lacquering, appointed sheet 6 Locker 7 appointed wall paper 8 interlock controller 9 finishing by appointed lacquering 10 sheet work(flower pattern)on black color glass 11 logo SCASI on counter H : 250 12 logo SCASI(person) on counter H : 10 13 appointed color glass 14 expansion 15 appointed moulding 16 black mirror, pattern sheet 17 finishing by appointed color glass / sus finish(indirect light-T5) 18 confectionery showcase : clear glass 19 beaded ice cream(730X670X860) 20 rock-climbing 21 terrace door

1 도장 마감 삼화 B-1013 2 기존건축 마감, 보호대 3 놀이기구 기둥, 보호대 4 도장마감, 벽화(포토대) 5 지정 도장 마감, 지정시트 6 락카장 7 지정벽지 8 연동제어기 9 지정 도장 마감 10 검정칼라 유리 위에 시트 작업(꽃문양) 11 카운터 위 로고 스카시 H : 250 12 카운터 위 로고 스카시(사람) H : 10 13 지정칼라 유리 14 전개 15 지정몰딩 16 흑경, 문양시트 17 지정칼라유리마감 / 서스마감(간접조명-T5) 18 제과쇼케이스 : 투명유리 19 구슬아이스크림(730X670X860) 20 암벽타기 21 테라스 도어

주방 그래픽 / kitchen graphic

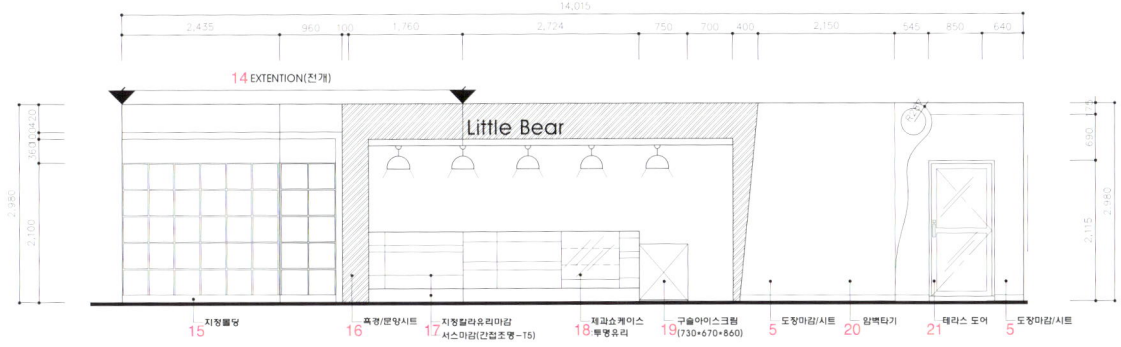

홀 입면도 B / hall elevation B

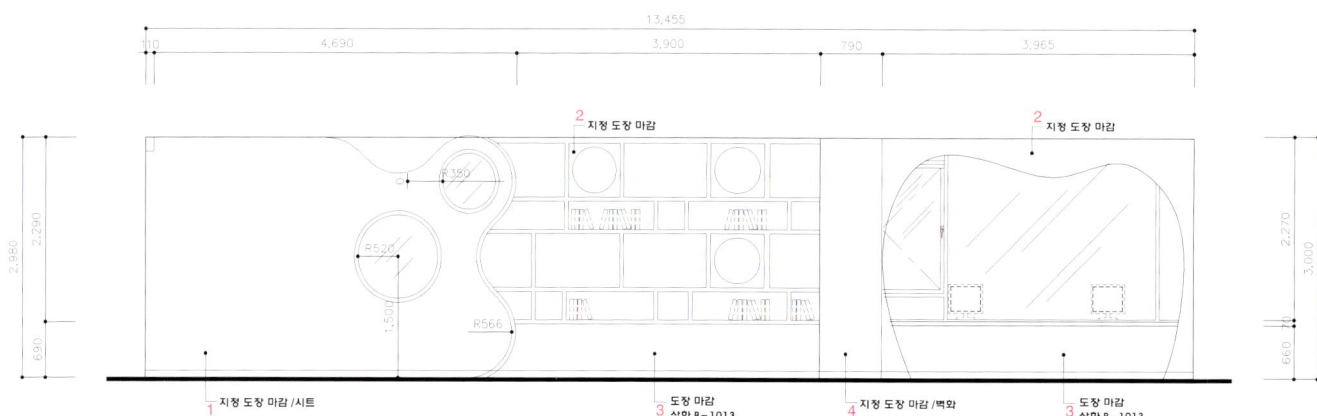

홀 입면도 C / hall elevation C

1 finishing by appointed lacquering, sheet 2 finishing by appointed lacquering 3 finishing lacquering samhwa B-1013 4 finishing by appointed lacquering, mural paintings 5 glass 6 SCASI(thickness 5mm) 7 T8 clear tempered glass / glass installing 8 T12 tempered glass / appointed sheet 9 Extention 10 finishing by lacquering samhwa B-1013 11 finishing by lacquering : maol sheet / finishing by lacquering : samhwa B-1013

1 지정 도장 마감, 시트 2 지정 도장 마감 3 도장 마감 삼화 B-1013 4 지정 도장 마감, 벽화 5 거울 6 스카시(두께 5mm) 7 T8 투명강화유리 / 거울 취부 8 T12 강화유리 / 지정시트 9 전개 10 도장 마감 삼화 B-1013 11 도장마감 : 마을 시트 / 도장 마감 : 삼화 B-1013

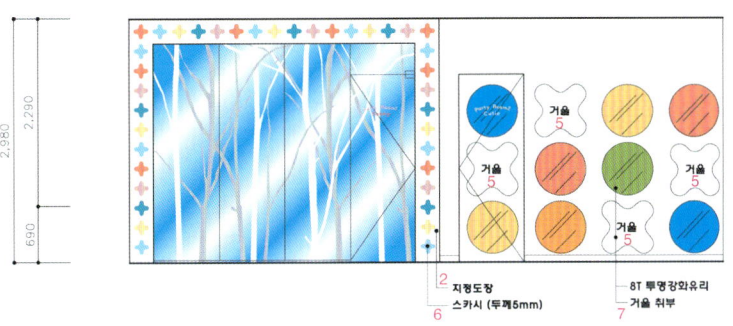

파티실 그래픽 / party room graphic

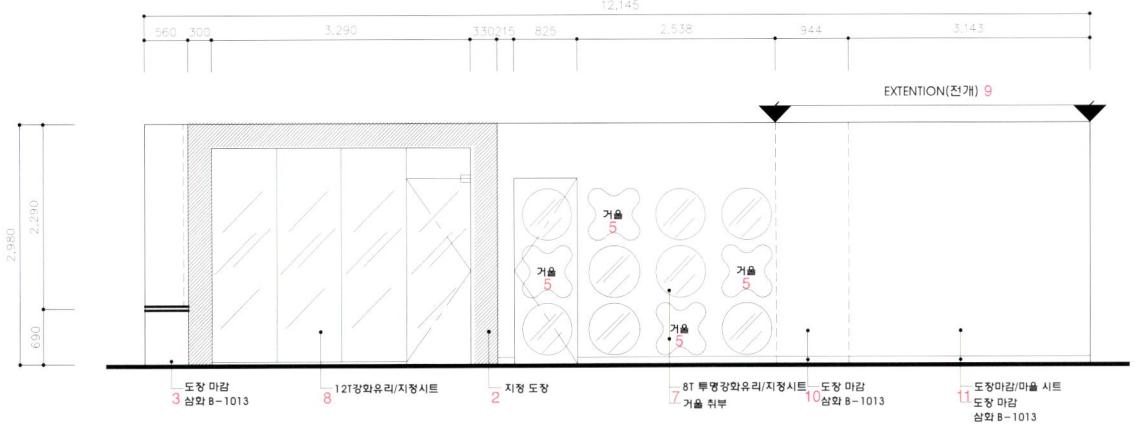

홀 입면도 D / hall elevation D

1 PDP 2 movable making chair 3 movable singing machine 4 finishing by lacquering samhwa B-1013 5 finishing by lacquering / sheet samhwa B-1013 6 T8 clear tempered glass 7 existing window / wood blind 8 T8 clear tempered glass / mirror installing 9 mirror 10 finishing by lacquering 11 sheet on finishing by lacquering 12 finishing by lacquering / sheet on finishing by lacquering 13 Solvent printing on glass

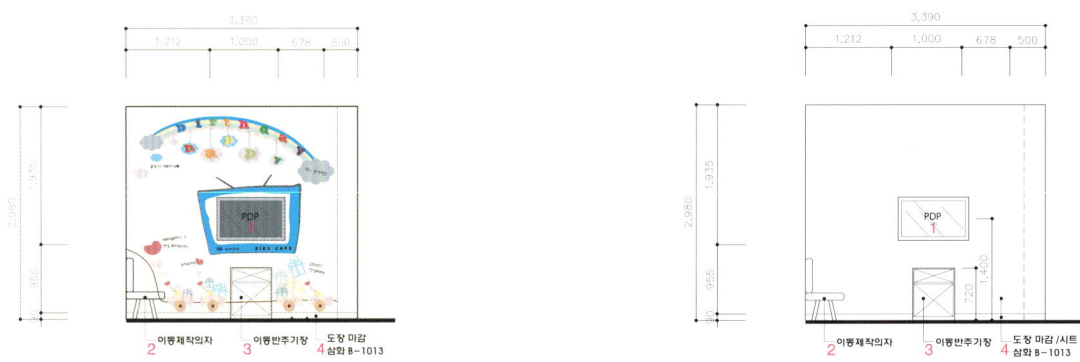

파티실 내벽 그래픽 / party room wall graphic

파티실 입면도 E / party room elevation

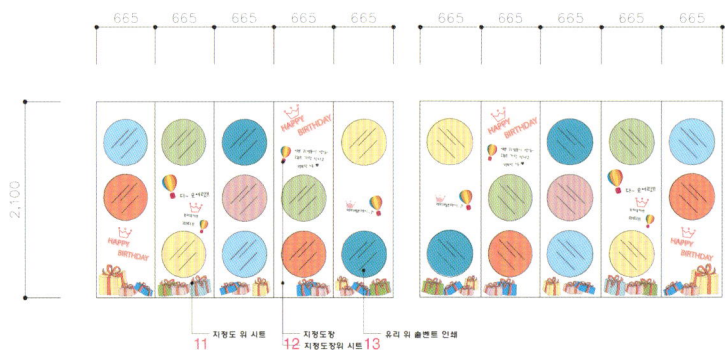

파티실 접이문 그래픽 / party room folding door graphic

1 BLACK MIRROR FINISHING 2 PAINTING 3 PAINTING / DOOR - BLACK FURNITURE / BIRCH PLYWOOD FINISHING ON BLACK PAINTING 4 BIRCH PLYWOOD FINISHING ON BLACK PAINTING 5 FILM 6 WALLPAPER 7 FURNITURE 8 WHITE LUMINOUS CEILING / GLAPHIC ON BLACK MIRROR / FILM 9 PAINTING / GRAPHIC ON GLASS 10 SILVER MIRROR 11 T12 TEMPERED GLASS

1 흑경 마감 2 도장 마감 3 도장 마감 / 도어 – 검은색 가구 마감 / 검은색 도장 위 자작나무 마감 4 검은색 도장 위 자작나무 마감 5 필름 마감 6 도배 마감 7 가구 마감 8 화이트 광천장 마감 / 흑경 위 그래픽 마감 / 필름 마감 9 도장 마감 / 유리 위 그래픽 마감 10 은경 마감 11 T12 강화 유리

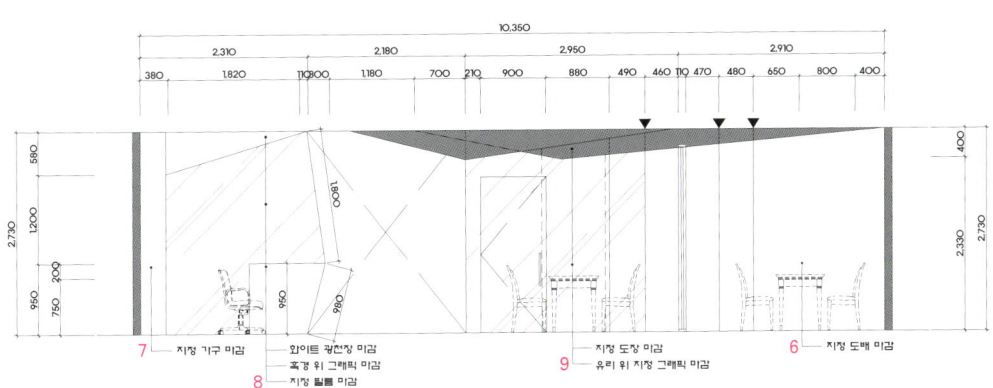

회의실 입면 C / meeting room elevation C

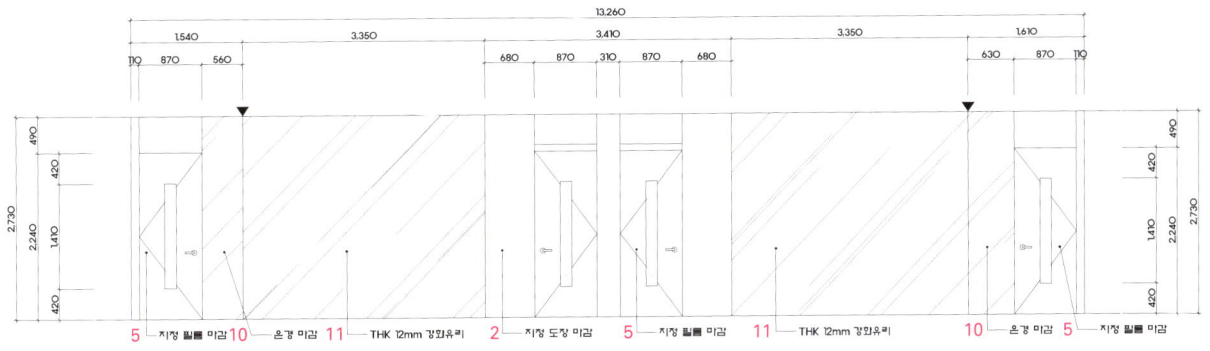

교실 입면 D / class elevation D

BAMBINI & KAGE

(주)예국디자인 | 이철수, 신현정 Yekukdesign | Lee Choul Soo, Shin Hyun Jung

호기심을 자극하다 유년시절 몸으로 체험한 공간에 대한 기억은 많은 영향과 변화를 가져온다. 생각의 폭과 깊이를 성장시키고, 삶의 방식과 태도에도 영향을 주기 때문에 아이들을 위한 공간에서는 정서적, 감성적 사고를 키우는 다양성과 변화가 요구된다.

숲 - 어린이의 눈동자에 빛을 담다.(BAMBINI 교육센터) 따뜻한 햇살과 비치는 물빛, 바람이 지나가는 숲을 디지털적인 시각으로 재해석하고 아이들의 호기심을 자극하는 컬러와 사고의 자유로움을 유도하는 그래픽으로 공간을 그렸다. 자연 그대로의 숲을 보여주는 것이 아니라 아이들이 스스로 공간을 통해 머릿속으로 상상하고 꿈꾸는 자신만의 숲을 그릴 수 있도록 하였다. 정서적 교감과 더불어 동적인 학습으로 즐거움을 배우는 BAMBINI 교육센터는 아이들의 눈동자에 빛을 담아주기에 충분한 공간이다.

숲 - 호기심이라는 렌즈로 숲을 보다. (KAGE 영재교육학술원) 호기심은 아이들이 가지는 최대의 장점이며 자기 계발에 꼭 필요한 바탕이다. 다양한 시각을 가진 아이들과 닮은 숲을 모티브로 여러 각도로 들여다보고 모듈화하였다. 숲의 색감을 담은 나무 조각들이 모여 아이들의 키를 훌쩍 넘는 나무가 되고, 나뭇잎 사이사이로 햇살이 떨어진다. 사선의 빛들과 수학적인 패턴들은 호기심이라는 렌즈를 통해 보는 아이들의 창의적인 감각을 높이는 역할을 한다.

Stimulates curiosity Memories of the spaces that we experienced during childhood greatly influence on the life. As it develops the width and depth of thinking while affecting the view and attitude for life, children need a space where can nourish them mentally and emotionally with many changes.

Forest - Contains the light in the eyes of children. (BAMBINI Education Center) We reinterpreted a forest with warm sunshine, transparent water, and fresh breeze in digital way and created the space in eye - catching colors and vibrant graphics. Instead of showing what a forest looks like, we decided to let them create their imaginary forests. BAMBINI Education Center is good enough for children who enjoy learning through emotional communication and dynamic activities.

Forest - See through a lens named curiosity. (Korea Academy of Gifted Education) Curiosity is the most precious nature of children and is necessary for self - development. We saw a forest in different aspects to meet requirements of children who have widely open eyes to the world, and we modulized the ideas. The wooden pieces in the colors of a forest become a tree which is taller than children's height and the tree filters the natural light of the sun through its leaves. The diagonal sunbeam and mathematic patterns promote children who see the world through a lens named curiosity to have creative thinking.

위치 경기도 화성시 반송동 218-1 계림계발빌딩 6층
용도 교육
면적 775m²
마감 바닥 - 데코 타일, 강화 마루 / 벽체 - 래커 도장, 은경, 동경, 백페인트 유리, 강화 유리, 모자이크 타일, 강마루 / 천장 - 비닐 페인트, 스트레치 천장 시스템
디자인팀 전미, 전구영, 이슬비, 김락진
그래픽팀 이동화
건축주 BAMBINI 교육센터 & KAGE 영재교육학술원
사진 (주)예국디자인 제공

Location 6F, Gyerimgyebal Building, 218-1, Bansong-dong, Hwaseong, Gyeonggi-do
Use Education
Area 775m²
Finishing Floor - Deco tile, Reinforced floor / Wall - Lacquer painting, Silver mirror, Bronze mirror, Back paint glass, Tempered glass, Mosaic tile, Tempered plywood floor / Ceiling - Vinyl paint, Stretch ceiling system
Photos offer Yekukdesign

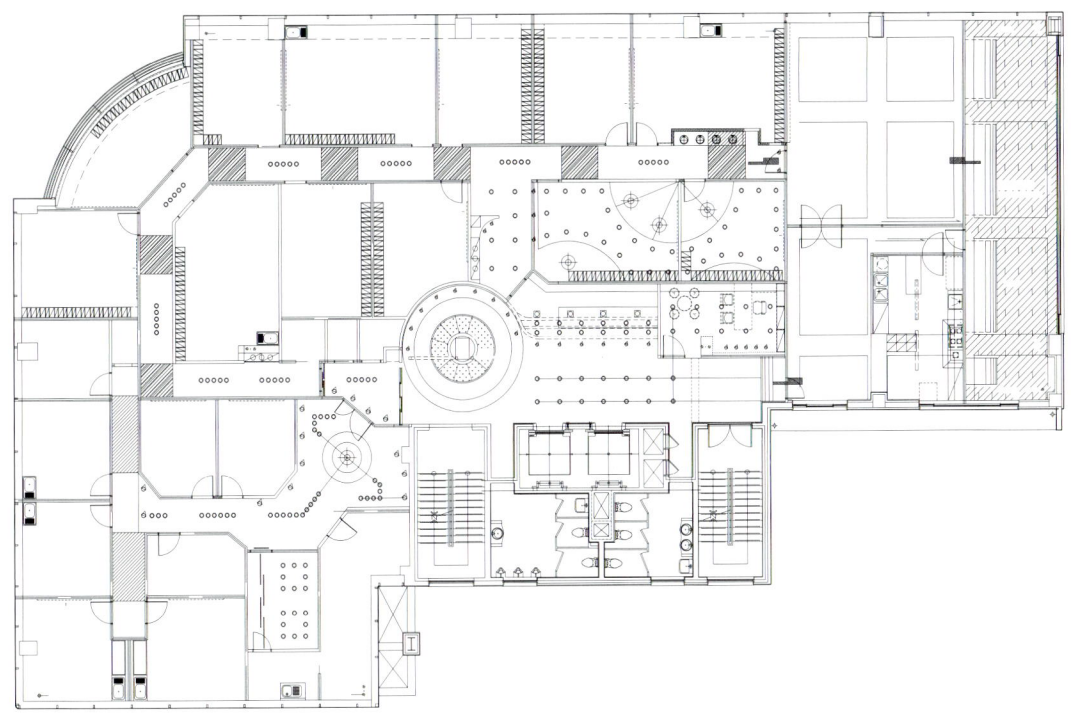

천장도 / ceiling plan

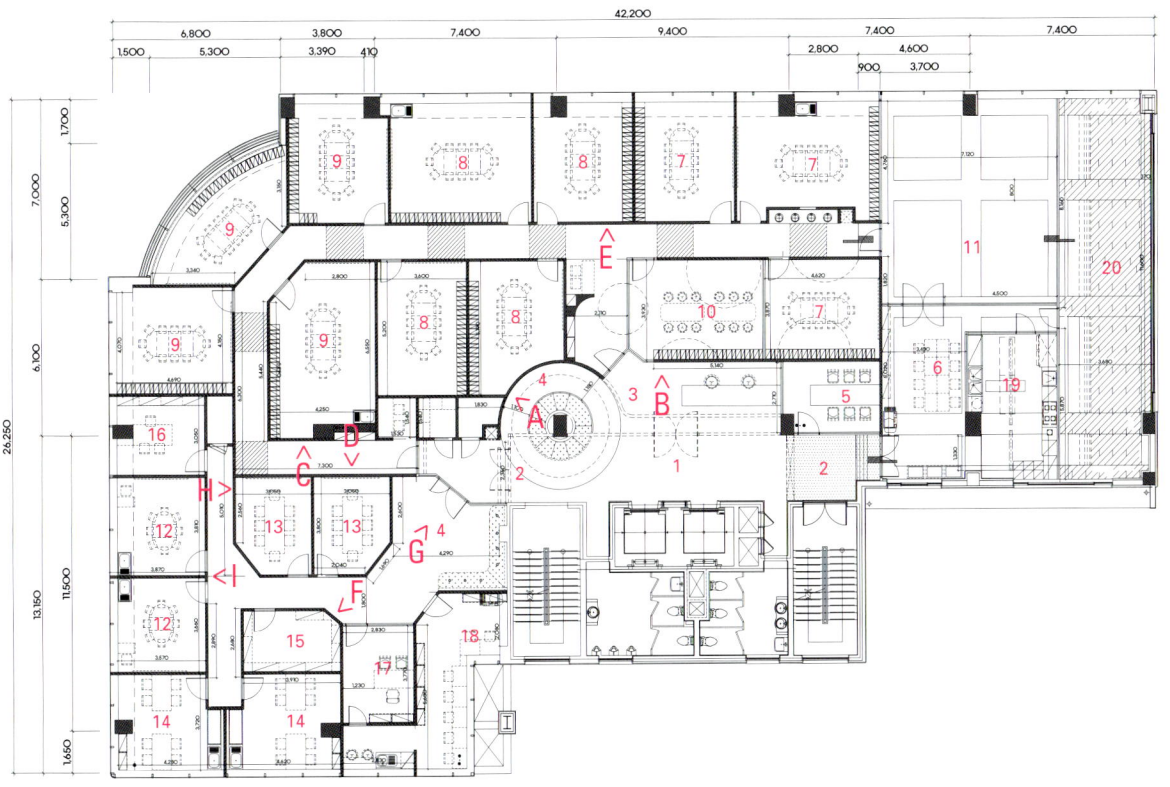

평면도 / floor plan

1 MAIN ENTERANCE 2 SUB ENTERANCE 3 BAMBINI, KAGE COMMON HALL 4 WAITING AREA 5 COUNSELING ROOM 6 TEACHER'S ROOM / CLASS PREPARATION ROOM 7 5-YEAR-OLD CLASSROOM 8 6-YEAR-OLD CLASSROOM 9 7-YEAR-OLD CLASSROOM 10 LIBRARY 11 GYM CLASS 12 INFANTS CLASSROOM 13 CHILDREN CLASSROOM 14 SCIENCE ROOM 15 DATA ROOM 16 TEST ROOM 17 DIRECTOR'S ROOM 18 TEACHER'S ROOM 19 KITCHEN 20 TERRACE

1 주출입구 2 부출입구 3 밤비니, 케이지 공용 홀 4 대기실 5 상담실 6 교무실 / 수업 준비실 7 5세 교실 8 6세 교실 9 7세 교실 10 도서실 11 체육실 12 영유아 교실 13 아동 교실 14 과학실 15 자료실 16 검사실 17 원장실 18 교무실 19 주방 20 테라스

1 BARRISOL 2 COLOR MDF 3 COLOR MDF LASER CUTTING 4 APP. PAINTING FIN. 5 APP. WALLPAPER FIN. 6 COLOR GLASS FIN. 7 SILVER MIRROR FIN. 8 BRONZE MIRROR FIN. 9 APP. PAINTING ON 2P FIN. 10 APP. COLOR GLASS FIN. ON 1P 11 APP. FILM FIN. ON MDF 12 GALVA APP. FILM FIN. 13 BASEBOARD(H : 80)

1 바리솔 2 컬러 MDF 3 컬러 MDF 레이저 커팅 4 지정 도장 마감 5 지정 벽지 마감 6 컬러 유리 마감 7 은경 마감 8 동경 마감 9 2P 마감 위 지정 도장 10 1P 위에 지정 컬러 유리 마감 11 MDF 위 지정 필름 마감 12 갈바 지정 필름 마감 13 걸레받이(H : 80)

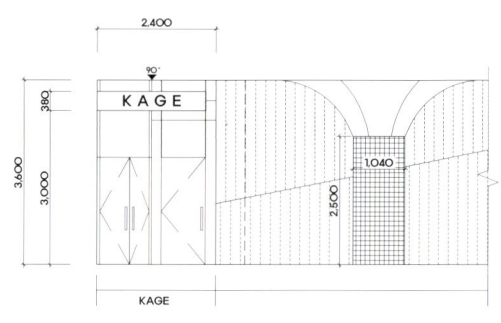

입구 입면 A / entrance elevation A

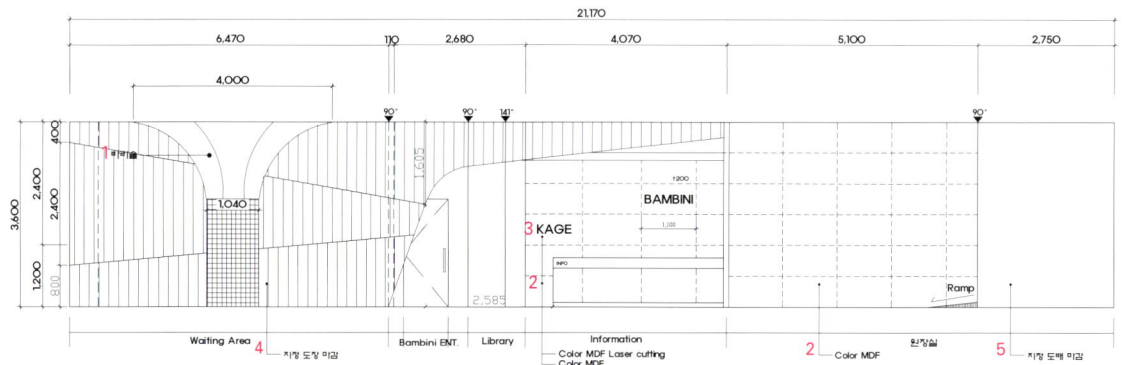

홀 입면 B / hall elevation B

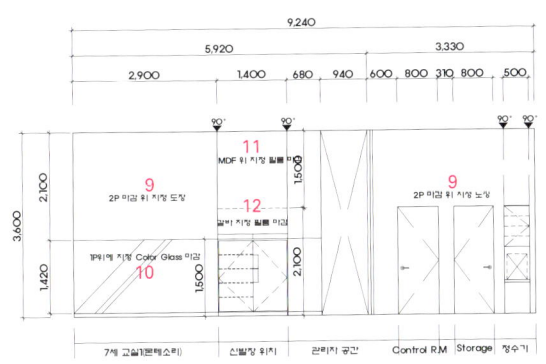

내부 입면 C / inner elevation C

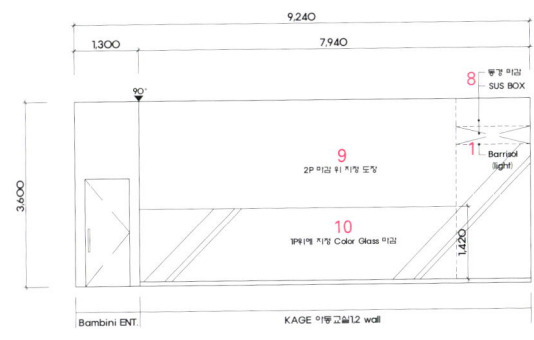

내부 입면 D / inner elevation D

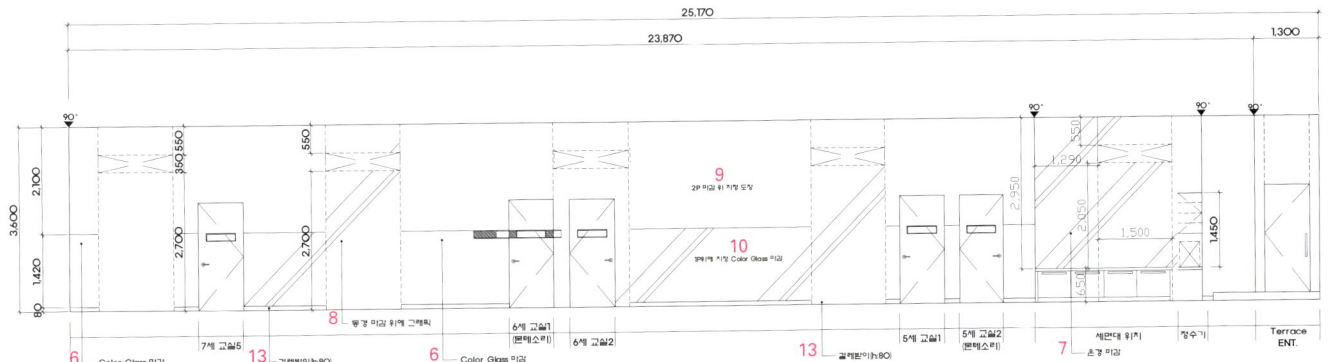

내부 입면 E / inner elevation E

KOREA ACADEMY OF GIFTED EDUCATION

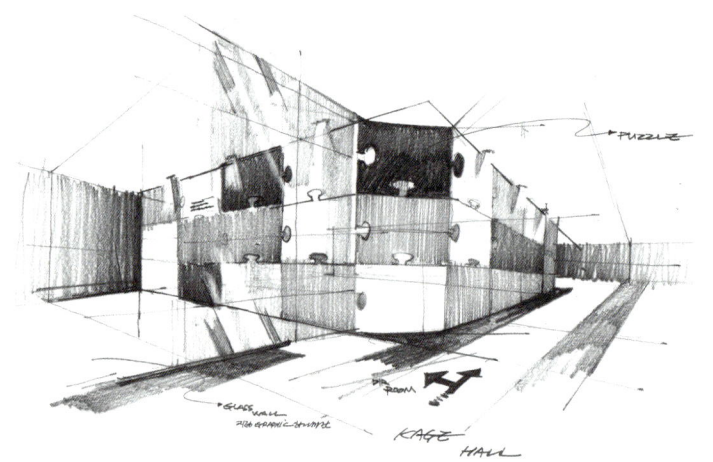

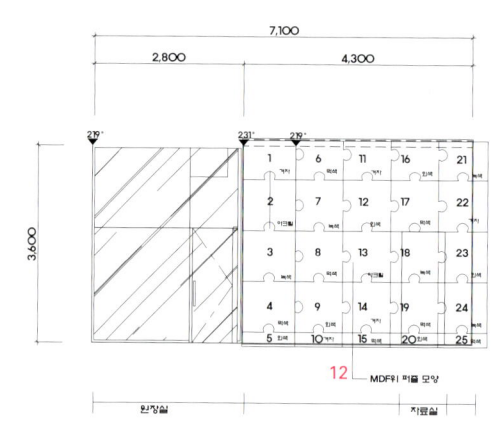

내부 입면 F / inner elevation F

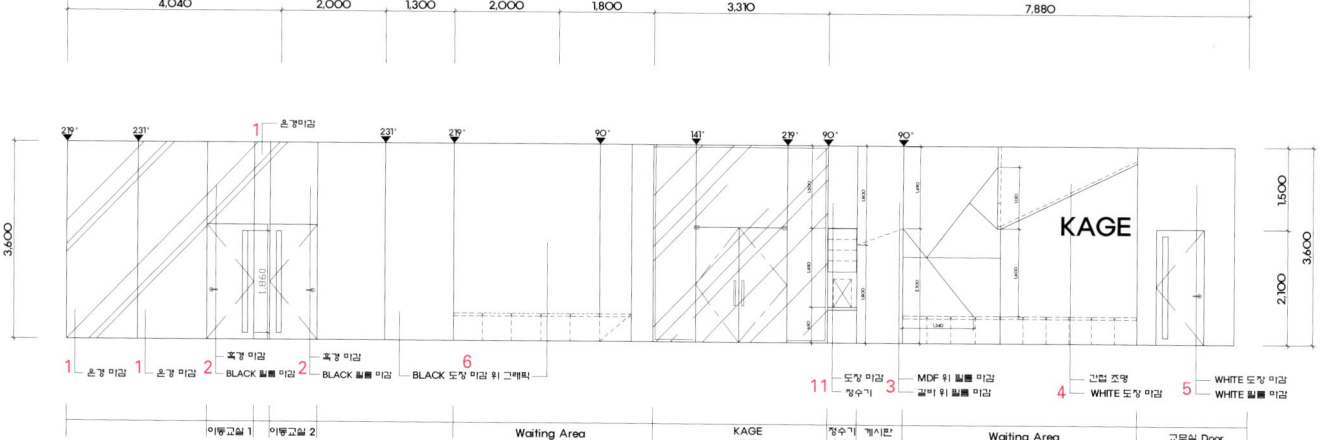

내부 입면 G / inner elevation G

1 SILVER MIRROR FIN. 2 BLACK MIRROR FIN. / BLACK FILM FIN. 3 FILM FIN. ON MDF / FILM FIN. ON GALVA 4 INDIRECT LIGHTING / WHITE PAINTING FIN. 5 WHITE PAINTING FIN. / WHITE FILM FIN. 6 GRAPHIC ON BLACK PAINTING FIN. 7 APP. PAINTING FIN. 8 WHITE FILM FIN. 9 BARRISOL / APP. PAINTING FIN. 10 APP. PAINTING FIN. / PAINTING FIN. ON MDF 11 PAINTING FIN. 12 PUZZLE SHAPE ON MDF

1 은경 마감 2 흑경 마감 / 검정 필름 마감 3 MDF 위 필름 마감 / 갈바 위 필름 마감 4 간접 조명 / 흰색 도장 마감 5 흰색 도장 마감 / 흰색 필름 마감 6 검정 도장 마감 위 그래픽 7 지정 도장 마감 8 흰색 필름 마감 9 바리솔 / 지정 도장 마감 10 지정 도장 마감 / MDF 위 도장 마감 11 도장 마감 12 MDF 위 퍼즐 모양

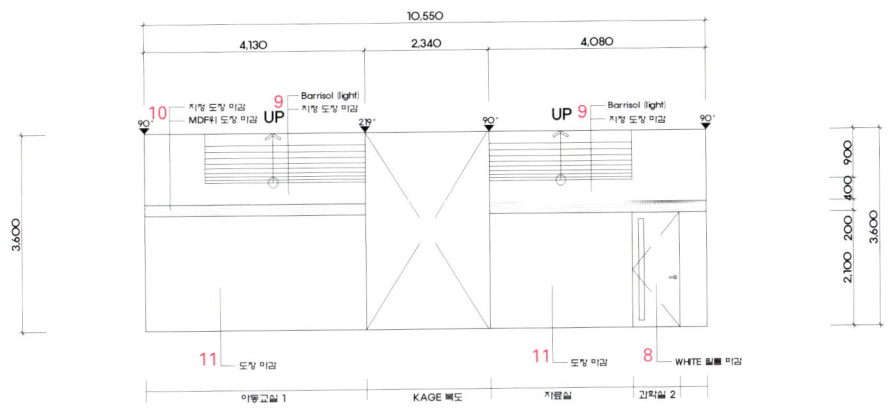

내부 입면 H / inner elevation H

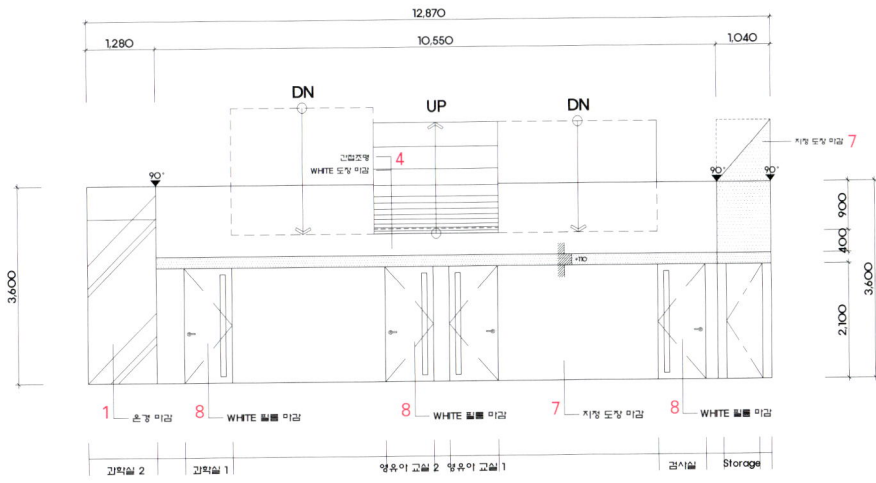

내부 입면 I / inner elevation I

발행	에이엔씨출판 주식회사
등록	제2004-000166호
발행인	정흥채
진행	출판기획부
출력·인쇄	삼성문화인쇄(주)
주소	서울특별시 강남구 테헤란로22길 15 에이엔씨빌딩 10층
전화	02-538-7333

ⓒ에이엔씨출판 주식회사
한국간행물 윤리위원회의 윤리강령 및 실천요강을 준수합니다.
본지에 게재된 내용을 사전 허가 없이 무단 복제 및 전재를 금합니다.

값 38,000원(38$)